Clinical Ecology

A New Medical Approach to Environmental Illness

clinical ecology

A New Medical Approach to Environmental Illness

Iris R Bell, MD, PhD

Common Knowledge Press
Bolinas, California

Common Knowledge Press is a subsidiary of Commonweal, a nonprofit institute engaged in service, research and demonstration programs in health, education and human ecology. Commonweal's major areas of program interest are: (1) Helping children with learning and behavior difficulties and their families; (2) investigating promising approaches to enhancing health and preventing disease; and (3) exploring environmental conditions that contribute to human health.

Additional copies of *Clinical Ecology* may be obtained by sending $4.95 plus $1.00 for postage and handling (California residents add sales tax) to Common Knowledge Press, P.O. Box 316, Bolinas, CA 94924.

Grateful acknowledgment is made to the following individuals for their thoughtful review of the manuscript:

Enoch Callaway, M.D.
Lawrence Dickey, M.D.
David S. King, Ph.D.
Sheldon Margen, M.D.
Joseph J. McGovern, M.D.
Penelope Post, M.A.
Theron G. Randolph, M.D.
Doris J. Rapp, M.D.
William J. Rea, M.D.
Phyllis L. Saifer, M.D., M.P.H.
P.B. Williams, Ph.D.

U.S. Library of Congress
Cataloging in Publication Data

Bell, Iris R.
CLINICAL ECOLOGY. A NEW MEDICAL APPROACH TO ENVIRONMENTAL ILLNESS

Includes bibliographical references.
1. Environmentally induced diseases.
1. Title.
[DNLM: 1. Ecology. 2. Environmental Health.
3. Hypersensitivity. WD 300 B433c]
RB152.B44 1982 616.9'8 82-7304
ISBN 0-943004-01-2

First printing July, 1982
Second printing October, 1982
Third printing September, 1983
Fourth printing March, 1986

Design by Jon Goodchild

Printed in the United States of America

But there are certain persons who cannot readily change their diet with impunity; and if they make any alteration in it for one day, or even for a part of a day, are greatly injured thereby. Such persons, provided they take dinner when it is not their wont, immediately become heavy and inactive, both in body and mind, and are weighed down with yawning, slumbering, and thirst; and if they take supper in addition, they are seized with flatulence, tormina, and diarrhoea, and to many this has been the commencement of a serious disease, when they have merely taken twice in a day the same food which they have been in the custom of taking once.

And thus, also, if one who has been accustomed to dine, and this rule agrees with him, should not dine at the accustomed hour, he will straightway feel great loss of strength, trembling, and want of spirits, the eyes of such a person will become more pallid, his urine thick and hot, his mouth bitter; his bowels will seem, as it were, to hang loose; he will suffer from vertigo, lowness of spirit, and inactivity—such are the effects; and if he should attempt to take at supper the same food which he was wont to partake of at dinner, it will appear insipid, and he will not be able to take it off; and these things, passing downwards with tormina and rumbling, burn up his bowels; he experiences insomnolency or troubled and disturbed dreams; and to many of them these symptoms are the commencement of some disease.

Hippocrates, *On Ancient Medicine (Adams, 1886).*

CONTENTS

clinical ecology

PREFACE

There has long been a need for a concise and authoritative overview of the field of clinical ecology that could serve as a reference for physicians, scientists, medical students, health care practitioners, patients and lay readers with a serious interest in this important and controversial field.

Clinical ecology is an alternative approach to the practice of environmental medicine. Practitioners of clinical ecology maintain that a broad range of common physical and psychological disorders can be triggered in susceptible individuals by chronic and often low-level exposures to foods, environmental chemicals, and natural inhalants.

This book was commissioned by the Commonweal Research Institute to fill the need for a comprehensive survey of clinical ecology. Its purpose is to describe the key concepts in clinical ecology, the historical development of the field, the relevant research literature, and current clinical practices. It was researched, written and produced with the support of the William H. Donner Foundation as a section of a forthcoming book of resources in clinical nutrition.

Michael Lerner, Ph.D.
President
Commonweal
Mac Arthur Prize Fellow
Institute for Health Policy Studies
University of California, San Francisco,
School of Medicine

INTRODUCTION

Clinical ecology has developed over the past 50 years as an interdisciplinary offshoot of environmental medicine. Clinical ecologists propose that chronic exposure to common foods, environmental chemicals, and natural inhalants—in addition to physical and psychosocial stressors—can trigger a wide range of mental, emotional, and physical disorders in susceptible individuals.[1,2] Although clinical ecology (CE) began within the field of allergy, the theories of the two disciplines and their methods for diagnosis and treatment have diverged significantly since the 1930's. Much controversy and debate over the field of CE has resulted. At present, the data suggest that environmental illnesses may be initiated by chronic exposure to low doses of multiple substances. Classical allergic mechanisms (i.e., antibody IgE) are not necessarily involved, but other dysfunctions of the immune system have been implicated.

Clinical ecology bases its procedures on two main hypotheses: (1) that the *total* load of low-dose environmental stressors is important in the induction of illness; and (2) that changes in the frequency of and intervals between exposures to a specific substance can mask the clinical manifestations of or alter the degree of sensitivity to that substance. In contrast to traditional allergists, clinical ecologists focus on psychiatric, central nervous system (CNS), and psychophysiologic syndromes which they believe are the results of (rather than the causes or correlates of) sensitivities to environmental agents. Generally, CE emphasizes the role of environmental chemicals and foods much more than that of natural inhalants (including dusts, pollens, animal danders, and molds) in the etiology of both psychiatric and physical disorders. The most frequently ingested or inhaled substances, including common foods and indoor as well as outdoor air pollutants, are considered by clinical ecologists to be the most common triggers of chronic ecological illnesses. Patients are frequently debilitated by their chronic symptoms and have frequently received other diagnoses, including "psychosomatic," by the time they come to treatment by a clinical ecologist. CE diagnosis depends largely on detection of specific sensitivities. Treatment involves avoidance, and/or symptom-neutralizing doses of diluted extracts of the offending agents, together with a rotating diet (which provides an interval of four or more days between exposures to safe foods) and an ecologically-sound oasis within the home.

A major challenge to clinical ecologists and researchers in environmental medicine is to test the fundamental clinical observations of CE with controlled studies and to explore mediating mechanisms within the body. The purpose of this survey is to summarize the key concepts of clinical ecology, to cite current research literature relevant to the main hypotheses, and to give an overview of diagnostic and therapeutic methods in CE.

CHAPTER ONE

Fundamental Concepts of Clinical Ecology

ACCORDING TO CLINICAL ECOLOGISTS, patients with environmental sensitivities usually have multiple symptoms involving many parts of the body. These symptoms can range from physical reactions such as a runny nose, abdominal bloating, headache, and fatigue, to psychological symptoms such as anxiety, depression, and difficulty in concentration. Each patient shows a characteristic set of problems. Table 1 lists some of the symptoms and syndromes to which environmental factors have been reported to contribute. The individual's tendency toward specific types of medical and/or psychiatric disorders is likely to be a complex function of age,[3] sex, heredity,[4,5] biological rhythms,[6] and nutritional status.[7,8]

For environmentally-sensitive individuals, there are reportedly numerous agents that can trigger symptoms. These substances include (a) chemical pollutants such as natural gas fumes from stoves and heating systems, formaldehyde in building materials and in permanent-press clothing, soft plastic odors from home furnishings and food storage containers, tobacco smoke, perfumes, chlorine and fluoride in tap water, food additives, and pesticides; (b) common foods such as milk, corn, wheat, egg, yeast, potato, and beef;

and (c) natural inhalants such as pollens, dusts, animal danders, and molds. Table 2 lists some of the agents which have been implicated in environmental illness.

Clinical ecology distinguishes itself in practice from other areas of environmental medicine such as allergy and toxicology by its greater emphasis on two main concepts: the role of total stress load and the role of time factors—including frequency of exposure and adaptation—in the expression of adverse reactions to environmental substances.

Total Load

The first hypothesis is that the onset of illness depends on the total stress load, including all of the psychosocial, physical, chemical, antigenic, or infective stressors that impinge on the individual (see Figure 1). Clinical ecologists suggest that low doses of substances which singly might be benign may interact additively or synergistically on some common pathways in the body to produce illness.[9,10]

Table 1. Reported Symptoms and Syndromes

1. *Skin:* Itching, burning, flushing, warmth, coldness, tingling, sweating behind neck, etc. Hives, blisters, blotches, red spots, "pimples."
2. *Ear, nose, throat:* Nasal congestion, sneezing, nasal itching, runny nose, postnasal drip. Sore, dry, or tickling throat, clearing throat, itching palate, hoarseness, hacking cough. Fullness, ringing, or popping of ears, earache, intermittent deafness, dizziness, imbalance.
3. *Eye:* Blurring of vision, pain in eyes, watery eyes, crossing of eyes, glare hurts eyes; eyelids twitching, itching, drooping, or swollen; redness and swelling of inner angle of lower lid.
4. *Respiratory:* Shortness of breath, wheeze, cough, mucus formation in bronchial tubes.
5. *Cardiovascular:* Pounding heart, increased pulse rate, skipped beats, flushing, pallor, warm, cold, tingling, redness or blueness of hands, faintness, precordial pain.
6. *Gastrointestinal:* Dryness of mouth, increased salivation, canker sores, stinging tongue, toothache, burping, retasting, heartburn, indigestion, nausea, vomiting, difficulty in swallowing, rumbling in abdomen, abdominal pain, cramps, diarrhea, itching or burning of rectum.
7. *Genitourinary:* Frequent, urgent, or painful urination; inability to control bladder; vaginal itching or discharge.
8. *Muscular:* Fatigue, generalized muscular weakness, muscle and joint pain, stiffness, soreness, chest pain, backache, neck muscle spasm, generalized spasticity.
9. *Nervous system:* Headache, migraine, compulsively sleepy, drowsy, groggy, slow, sluggish, dull, depressed, serious, crying; tense, anxious, stimulated, overactive, restless, jittery, convulsive, head feels full or enlarged, floating, silly, giggling, laughing, inebriated, unable to concentrate, feeling of separateness or apartness from others, amnesia for words or numbers or names, stammering or stuttering speech.

SOURCE: Miller, J.B. *Food Allergy: Provocative Testing and Injection Therapy.* Springfield: Charles C Thomas, Publisher, 1972, p. 21.

For example, if the only stressor in a patient's total environment were milk, he or she would probably tolerate it well. Most patients, however, have multiple sensitivities and are frequently subjected to many of their triggering stressors at the same time. A patient might drink milk while eating a full meal of 6-10 other offending foods, breathing tree pollens and natural gas, cooking with chlorinated and fluoridated water, and feeling upset by financial problems.

Clinical ecologists believe that illness appears when the total stress load ex-

Table 2. Chemical Stressors in Homes

Aerosols
Air deodorizers, sprays
Aluminum pots and pans
Ammonia

Bleaches

Car exhaust fumes entering the house from an open window or an attached garage
Cedar-lined closets
Charcoal
Chlorinated or fluoridated water
Christmas tree needles (which contain resin)
Cosmetics

Deodorants, anti-perspirants
Detergents
Disinfectants
Dyes

Electric blankets (because of plastic wires)

Felt-tip pens
Flameproof mattresses
Floor cleaners and waxes
Food additives
Formaldehyde*
Fungicide-treated wallpaper
Furniture polish

Gas stoves and other gas appliances

Hair sprays
Heat-sealed soft plastic packages

Insecticide sprays and no-pest strips

Kerosene

Lacquer

Medications
Mineral oil
Mothballs and moth crystals
Mothproofed shelf paper
Mouthwash

Nail polish
Newspaper print (inks and solvents)

Oils for fans, sewing machines, etc.
Oven cleaner

Paint fumes
Paraffin
Perfumes, aftershaves
Permanent-press clothing
Pesticides
Pillowcases and sheets of synthetics
Pine-scented cleaners
Plastics (mattress covers, tablecloths, shower curtains, draperies, usually vinyl, food wrappers, shelf paper)

Refrigerant gas (check for leaks)
Rubber-backed carpets
Rubbing alcohol

Scented soaps
Shampoos
Soft plastic food containers and wraps
Solvents used for duplicating machines
Smoke from frying foods
Sponge rubber (mattresses and upholstery)
Stainproof upholstery and carpets
Synthetic clothing

Teflon pots and pans
Tin cans with phenol lining
Tobacco smoke
Toothpaste
Turpentine

Varnishes containing pesticides
Varsol

SOURCE: Golos, N., Golbitz, F.G., Leighton, F.S. *Coping with Your Allergies.* New York: Simon and Schuster, 1979, pp. 46-47.

**Added by I.R. Bell*

ceeds some threshold for the patient and that recovery can occur when the total load is reduced below that threshold. As a result, they attempt a comprehensive investigation of the possible environmental incitants to illness in order to reduce the total stress load enough to permit recovery of health.

No direct experimental tests have been done on the validity of the total-load hypothesis. There are, however, certain epidemiological data consistent with that hypothesis. For example, painters occupationally exposed to mixtures of organic solvents exhibited significantly more nervous system signs and symptoms than did unexposed control workers.[11] Notably, the total exposure to any one of the solvents was below the accepted threshold for adverse effects.[11,12] One explanation of these findings consistent with the hypotheses of CE is that individual chemicals interacted with one another to produce effects that none alone would cause at low dose.[13,14] Of course, the retrospective nature of these studies also leaves the data open to other interpretations.[15]

In contrast to the clinical ecologists who emphasize the total-load concept, other researchers have recently focused, with variable success, on a one item/one illness model of environmentally-induced illness. Sensitivity to wheat gluten, for example has been associated with schizophrenia,[16-19] and adverse reactions to food additives[20-24] and to sugar[25] have been implicated in childhood hyperkinesis. Clinical ecologists, however, assert that there is usually no one-to-one correspondence between a single food or chemical and a particular diagnosis.

Figure 1. Model of Environmental Illness

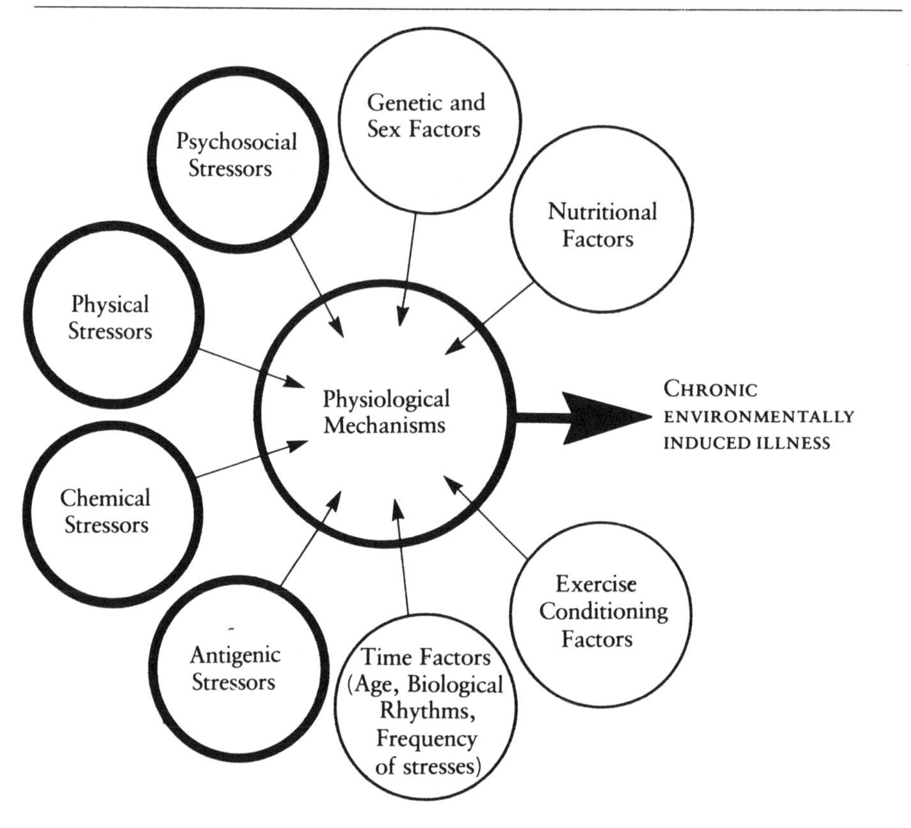

They claim that without the individualized and comprehensive approach of CE, clinicians and researchers will fail to detect all of the patients in whom ecologic factors are a significant health problem and all of the environmental substances which trigger illness in those patients.

Adaptation

The second major concept in CE is that of adaptation.[9] Here the *intervals* between exposures to offending foods are important. According to clinical ecologists, the greatest response to many stressors is usually evoked by the initial exposure after sensitization has developed. If the stressor is repeated often enough, however, the body adapts by progressively dampening its acute response to each subsequent exposure. Eventually, the patient adjusts to the extent that function appears almost normal after recurring exposures to the stressor.

A common example of this phenomenon is the response to tobacco smoke. A person may cough, choke and feel dizzy with the first cigarette but have minimal acute symptoms with the 70,000th cigarette after 10 years of daily smoking. A chronic illness such as lung cancer may be developing insidiously, although the marked acute symptoms after each cigarette may be greatly diminished. Similarly, in clinical ecology, a patient with environmentally-induced rheumatoid arthritis might originally notice acute flares of joint pains and lethargy within hours after eating an occasional hamburger. If the patient begins to eat beef daily, he or she may develop rheumatoid arthritis after a few months. Chronic joint stiffness and pain, as well as joint deformities, would have no obvious relationship in time to a particular meal containing beef.

Thus recurrent exposures to non-tolerated items may induce adaptation that would mask the acute worsening of symptoms after any given exposure, while still permitting chronic damage, with low-grade, dampened symptoms.[26] Clinically, this adaptation process means that patients and physicians may fail to suspect some environmental substances in the etiology of various chronic syndromes.[27]

A number of studies[28-31] have examined adaptation to ozone, an air pollutant that occurs both outdoors and indoors.[32-37] These studies demonstrated in the laboratory that human subjects with a history of respiratory problems exhibited adaptation to repeated ozone exposures within a period of only four days.[28] Initially, the subjects showed a decline in lung function on exposure to the ozone but, by the fourth day, they had recovered toward their baseline lung function. The idea for this research developed from the observation that when ozone levels are extremely high, residents of smoggy cities (who are presumably adapted to ozone) report fewer acute respiratory illnesses than do visitors from less polluted cities.[29,36,38]

Frequency of Exposure

According to clinical ecologists, frequent exposure can induce or increase sensitivity to a given substance,[39] while avoidance or lessened frequency of exposure can reduce the degree of sensitivity. As noted above, however, increased frequency of exposure is thought to induce not only greater sensitivity, but also adaptation. Adaptation dulls acute responses but causes low-grade chronic problems. For example, an individual might tolerate corn well until the age of 35 without acute or chronic symptoms. Then, with age and a high stress load, he might develop a sensitivity to corn while eating it often in various forms. Because the sensitivity developed while he was ingesting corn on a daily basis, he simultaneously develops an adaptation to corn and consequently does not experience acute symptom flares. Instead, he begins to suffer from chronic dull headaches all day, everyday, and never associates them with a clear-cut cause. Thus, adaptation masks the onset of chronic sensitivity to the offending food. Direct experimental tests of these hypotheses have not yet been done,[40,41] although clinical reports of such phenomena are extensive.[1,2,9,39,42-49]

CHAPTER TWO

Historical Development of Clinical Ecology

THE FIELD OF CLINICAL ECOLOGY originated in the first half of the 1900's with the observations of a number of American and British internist-allergists. One of the earliest books on the subject, *The Food Factor in Disease,* was written in 1905 by the Australian physician, Francis Hare.[50] Hare and later pioneers, such as American allergist Albert Rowe,[51,52] noticed the importance of *elimination of specific foods* from the diet in aiding recovery from a variety of syndromes such as ulcerative colitis, regional enteritis, bronchial asthma, eczema, migraine headache, and the nephrotic syndrome. Their observations led to the notion that the ability of a particular food to set off illness does not necessarily correspond to its nutritional content. Although two foods may have comparable carbohydrate, protein, and fat contents, one food might trigger symptoms in a given individual while the other will not. For example, wheat flour may cause joint pain in a patient with rheumatoid arthritis, but another carbohydrate source, such as arrowroot, may be tolerated.

Challenge Feeding Tests

Another American allergist, Herbert Rinkel, proposed the major hypothesis that intervals between exposures to an offending food are crucial in determining whether or not that food will cause acute symptoms. Rinkel made the surprising observation that many patients with chronic symptoms will actually experience a temporary flare in their illness in the first 3-4 days during which they avoid an offending food. Furthermore, eating that food during the symptom flare often leads to relief of the symptoms.[39,53] Rinkel's allergist colleague, Theron Randolph, termed this curious phenomenon "food addiction,"[54] due to (a) the withdrawal-like increase in symptoms; (b) the temporary relief of symptoms on re-exposure to the offending item; and (c) the reports of psychological craving for offending foods.[55] If patients continued to avoid the food for more than the three or four day period, their symptoms would spontaneously improve. Interestingly enough, after 3-4 days of avoidance, eating the food would set off rather than relieve the patient's symptoms. The symptoms often would be more exaggerated in number and severity than those in the chronic state.

Although experimental verification of the food-addiction hypothesis has not been obtained, the plausibility of this phenomenon has increased in recent years with the discovery that certain substances associated with the digestive process have opiate-like activity.[56-63] Digestion of wheat gluten produces peptides with an opiate-like effect.[56] A similar effect is induced by various endogenous peptide hormones such as cholecystokinin,[57] substance P,[58] and bradykinin,[62,63] which can be activated during digestion. The possible physiological mechanisms of food addictions are still a largely unexplored area for future investigation.

Rinkel utilized his clinical observations about the role of timing in food sensitivities in developing the challenge feeding test for diagnosis of food sensitivities. This procedure involved deliberate avoidance of a suspected food for four days, followed by a challenge meal composed entirely of that food on the fifth day. An adverse acute reaction to an offending food would be most intense at that time and thus would facilitate detection of a sensitivity. The challenge feeding test is still the standard with which newer tests for determining adverse reactions to foods are compared.

Because the challenge test intensifies acute reactions, it can unmask sensitivities to substances to which the body has previously adapted. Avoidance of foods which trigger acute attacks on challenge feeding tests leads to remission of chronic syndromes, according to clinicians who use this technique. The development of the challenge test provided a basis for deciding which foods to eliminate from each patient's diet. In the earlier elimination programs, the physician instructed the patient to avoid all foods that appeared in standard lists of probable offenders, without testing to determine which foods actually caused problems for that individual.[52]

Rinkel took his work on the role of timing in diet-induced illnesses one step further. He observed that the more often a patient ate a particular food, the more likely he or she would be to develop a sensitivity to it. Conversely, as a patient avoided a food, the degree of sensitivity to it decreased over time. After a 3-6 month avoidance period, the degree of sensitivity to an offending food was usually low enough that it could be eaten infrequently without the reactivation of acute or chronic symptoms. Once sensitivity existed, however, the patient could never again tolerate the food daily or even more often than once in four days. Frequent use seemed to reactivate the sensitivity and the re-emergence of symptoms, generally within 2-3 weeks. Consequently, Rinkel developed a therapeutic schedule, the rotation diet, to minimize the risk of developing sensitivity to a given food.[64] With a rotary diet, each safe or non-offending food is eaten not more than once every four or more days. Exposures to biologically-related foods such as wheat and rye also must be separated, preferably by intervals of four or more days. The intervals between re-exposures reduce the risk of developing sensitivity in the first place. This schedule also makes each meal on a rotary diet into a challenge test for unmasking any sensitivity as it might develop.

The role of timing in the ability of most environmental agents to harm the body has had little experimental study. Nevertheless, some data suggest that certain toxic agents may do more damage when exposure is intermittent than when it is continuous. One research group has found in experiments on animals that a given dose of nitrogen dioxide is more toxic on an intermittent than on a continuous exposure schedule, while the opposite is true for benzene.[40,41] These observations raise questions about the potential risks and benefits of maintaining patients on a long-term program of avoidance and intermittent rotation, rather than leaving them on their usual pattern of continuous exposure to various offending foods and environmental chemicals. If the avoidance is complete, then the risk of adverse effects of a chemical is obviously irrelevant. On the other hand, clinical ecology patients may avoid many foods and chemicals at home most of the time only to encounter unintentional, hence intermittent, exposures to those substances on venturing out of their safe home environments into traffic fumes, offices, stores, schools, and the homes of others. If such exposures are sufficiently frequent and susceptibility fails to decrease, it is theoretically possible that more, rather than less, damage to health could sometimes result, depending on the nature of the substances encountered. Generally clinical ecologists report that most patients in their program achieve improved health. Nonetheless, the need for prospective and retrospective studies on the question of timing for specific foods and chemicals is evident.

Environmental Chemical Incitants

Awareness of the problem of environmentally-induced illness is growing.[65] Theron Randolph, one of the founders of clinical ecology, was the first to suggest that for a given patient environmental chemicals could cause the same symptoms as could foods.[1] In his view, air pollution in the home, school, and office may pose an even greater public health problem than do occupational chemical exposures in factories and on farms. Because the number of potential offending substances is so large, Randolph saw a need for comprehensive challenge testing and treatment.[66] His goal was to continue testing and eliminating offenders until the patient achieved maximum clinical improvement.

For this purpose, Randolph originated the diagnostic procedure of hospitalization in an environmental unit. There he was able to remove very ill patients from all of their chronic chemical exposures, fast them on spring-water for 4-5 days, test them with single-food challenge meals until a basic rotary diet of safe foods could be established, and do selected chemical challenge tests as well.

There are now several hospitals around the U.S. with environmental diagnostic units to which patients from all over the world can go for a period of two to four weeks. These units are constructed with materials and furnishings that have minimal chemical outgassing. Special filter systems separate the air in the unit from that in the rest of the hospital.[67] Patients, staff, and visitors must wear natural, untreated fibers such as cotton and avoid the use of certain soaps, hairsprays, and perfumes which would contaminate the ward environment. The test meals are prepared from foods that contain the least possible chemical contamination. Patients receive extensive education. They learn about the nature of environmental illness, about methods of self-testing for possible offending foods and chemicals at home, and about managing their diet and environment after discharge from the hospital. With this approach, clinical ecologists report that in weeks or even days they can effect remissions in many patients with chronic illnesses.

Clinical Ecology within the Context of Medicine

Although it would appear to have much significance for the practice of medicine and psychiatry, clinical ecology has had surprisingly little impact to date on those disciplines. The reasons for this are complex. They include CE's lack of rigorous documentation, the anecdotal nature of the findings, and the interdisciplinary nature of the work. The latter means that CE lacks a single traditional field which could accommodate it and test its basic concepts.

The CE reports of central nervous system, psychophysiologic, and psychiatric syndromes have helped push it outside the boundaries of allergy, which has a more narrow clinical focus on immunoglobulin E-mediated events, associated mediator release (e.g., histamine), and consequent local physical symptoms. Moreover, although many aspects of CE are related to toxicology, clinical ecologists disagree with some fundamental premises in that field. In the first place, clinical ecology suggests that many substances (such as common foods) that are presumed non-toxic under usual conditions are potential triggers of illness. Clinical toxicologists are generally concerned with substances that are more consistently toxic. Furthermore, in contrast to many toxicologists, clinical ecologists state that safe thresholds for exposure to any given environmental agent may not exist. Clinical ecologists suggest that low doses of substances, especially in combination with one another as part of the total load, may be harmful to larger segments of the population than many toxicologists recognize.[65,68]

Timing may have been another reason why clinical ecology did not have more impact on the field of psychiatry. Randolph's original reports of a cause-effect relationship between foods and chemicals and psychiatric symptoms appeared in the 1950's,[69] when modern psychopharmacology was expanding and the use of phenothiazine tranquilizers was becoming widespread. Only recently have researchers begun to explore the possible food and chemical factors in mental and emotional illnesses.[11,12,16-25,70,71]

CHAPTER THREE

Symptom Patterns in Environmental Illness

One of Randolph's conceptual contributions has been his hypothesis that, in a given patient, the physical and psychological symptoms triggered by environmental agents generally follow a characteristic pattern. He has described this pattern as a biphasic continuum of reaction levels.[9] Figure 2 and Table 3 illustrate the model of "stimulated" and "withdrawal" phases which many patients are observed to experience during acute reactions.

According to Randolph, patients often begin an adverse reaction to a food or chemical with an initial stage of stimulation during which the level of excitation ranges from +1 to +4. This stage is characterized by an increase in behavioral and psychological activity, ranging from slight jitteriness to mania. In the course of a given reaction, the patient may then pass through a stage of withdrawal in which central nervous system function is depressed and the level of stimulation drops to a −1 to a −4 level. Withdrawal manifestations may include depression, sleepiness, or fatigue. Physical symptoms often emerge as the patient is recovering from a reaction. Local allergy such as hives, runny nose, asthma (level −1), and other physical symptoms such as headache, muscle and joint pains (level −2) represent less intense reactions than do the psychological symptoms. As the patient recovers from an acute reaction, he or she oscillates up and down through progressively less severe levels of stimulation and depression of function before returning to baseline. A single acute reaction can take from a few hours to a few days to subside. The clinical state of a particular patient is some net result of acute and chronic reactions to multiple items, all in different phases of the time course for reactions.

For instance, a patient in an acute reaction to beef might exhibit increased motor activity, inappropriate laughter and euphoria (level +3), followed by sleepiness and fatigue (level −2). On awakening from a nap, this patient might have a runny nose (level −1), but a normal mood. After a few more hours, he might be back to normal, symptom-free status (level 0).

Randolph and others before him[72] have also noted in some patients an apparent alternation between physical (−1) and psychotic (−4) levels of symptomatology. That is, some patients will be free of asthma while delusional or hallucinating only to redevelop asthma attacks as they recover their sanity. A negative correlation between the occurrence of local allergies or rheumatoid arthritis and the occurrence of psychosis in the same patient has long been controversial.[73,74] However, the possibility of this phenomenon in a subgroup of individuals with both physical and psychiatric diagnoses remains an important area for investigation.

A number of observers[70, 75-78] have reported a specific symptom complex in

Figure 2. Symptom Progression of a Single Reaction

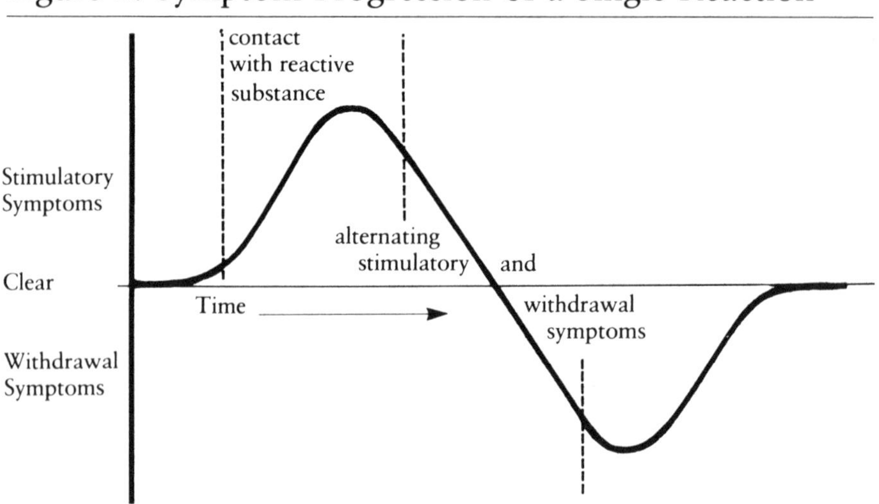

Source: O'Banion, D.R. *Ecological and Nutritional Treatment of Health Disorders*. Springfield, Illinois: Charles C Thomas, Publisher, 1981, p. 68.

Symptoms progress from dominant stimulatory to dominant withdrawal, with alternating stimulatory and withdrawal symptoms during the transition period between the two dominant phases. See also Table 3 for types of symptoms at each level.

Table 3. Stimulatory and Withdrawal Levels of Symptoms in Adverse Ecologic Reactions

	Directions:		Start at zero (o). Read up for predominantly Stimulatory Levels. Read down for predominantly Withdrawal Levels.
STIMULATORY LEVELS	+ + + +	**Manic with or without convulsions**	Distraught, excited, agitated, enraged and panicky. Circuitous or one-track thoughts, muscle twitching and jerking of extremities, convulsive seizures and altered consciousness may develop.
	+ + +	**Hypomanic, toxic, anxious and egocentric**	Aggressive, loquacious, clumsy, (ataxic), anxious, fearful and apprehensive; alternating chills and flushing, ravenous hunger, excessive thirst. Giggling or pathological laughter may occur.
	+ +	**Hyperactive, irritable, hungry and thirsty**	Tense, jittery, hopped up, talkative, argumentative, sensitive, overly responsive, self-centered, hungry and thirsty, flushing, sweating and chilling may occur as well as insomnia, alcoholism and obesity.
	+	**Stimulated but relatively symptom-free (sub-clinical)**	Active, alert, lively, responsive, and enthusiastic with unimpaired ambition, energy, initiative and wit. Considerate of the views and actions of others. This usually comes to be regarded as "normal" behavior.
NORMAL HOMEOSTASIS	o	**Behavior on an even keel as in homeostasis**	Children expect this from their parents and teachers. Parents and teachers expect this from their children. We all expect this from our associates.
WITHDRAWAL LEVELS	-	**Localized allergic manifestations**	Running or stuffy nose, clearing throat, coughing, wheezing (asthma), itching (eczema or hives), gas, diarrhea, constipation (colitis), urgency and frequency of urination and various eye and ear syndromes.
	- -	**Systemic allergic manifestations**	Tired, dopey, somnolent, mildly depressed, edematous with painful syndromes (headache, neckache, backache, neuralgia, myalgia, myositis, arthralgia, arthritis, arteritis, chest pain) and cardiovascular effects.†
	- - -	**Depression and disturbed mentation**	Confused, indecisive, moody, sad, sullen, withdrawn or apathetic. Emotional instability and impaired attention, concentration, comprehension and thought processes (aphasia, mental lapse and blackouts).
	- - - -	**Severe depression with or without altered consciousness**	Nonresponsive, lethargic, stuporous, disoriented, melancholic, incontinent, regressive thinking, paranoid orientations, delusions, hallucinations, sometimes amnesia, and finally comatose.

SOURCE: Randolph, T.G., Specific adaptation. *Annals of Allergy.* 40:333-45, 1978.

† *Marked pulse changes or skipped beats may occur at any level.*

many environmentally-ill patients. This syndrome includes fatigue, muscle and joint aches, drowsiness, difficulty in concentration, nervousness, and depression (see Table 4). Women reportedly outnumber men with this problem in a ratio of 2.5 to 1.[75] Allergists who first saw these symptoms termed them the "allergic toxemia"[75] or "allergic tension-fatigue" syndrome.[77] This syndrome, which is descriptively similar to the functional psychiatric diagnosis of neurasthenia,[79] has also been noted in patients without classic allergies who nonetheless have chronic adverse reactions to foods and environmental chemicals.[76,80] Although there has been little controlled testing of whether neurasthenia is often caused by environmental illness, some epidemiological studies are consistent with a relationship between this pattern of psychiatric symptoms and chemical environmental factors. For example, some studies show that male workers exposed to diesel jet fuels report the symptoms of neurasthenia such as fatigue, depression, dyspnea, dizziness, palpitations, and sleep disturbances significantly more often than do a control group of workers who have not been exposed to such chemicals.[70,80]

Because environmental illnesses are apparently characterized by patterns of multiple symptoms in many parts of the body, including the CNS, it is understandable that such patients have often received diagnoses of neurasthenia,

Table 4. Symptoms of "Allergic Tension-Fatigue Syndrome"

Symptom	*Percentage of Patients (N = 70) (no control group reported)*
Fatigue	94%
Food sensitivities	91
Gastrointestinal symptoms	63
Headache	50
Arthralgias	47
Drowsiness	37
Myalgias	36
Nervous tension	34
Nasal symptoms	34
Depression	30
Difficulty concentrating	27
Irritability	20
Confusion	17
Hives	16
Insomnia	11
Aching in chest	10
Fever	10
Eczema	10
Tachycardia	10

Adapted from Rowe, A.H., and Rowe, A., Jr. *Food Allergy. Its Manifestations and Control and Elimination Diets.* Springfield, Illinois: Charles C Thomas, Publisher, 1972, p. 357.

Patient sample reportedly included 72% female and 28% male patients, with mean age of 42 years (range 10-70).

hysteria, somatization disorder, and various other psychosomatic disorders. Some of these patients may be hysterics with psychogenic symptoms. It is also possible that certain patients have a way of coping with symptoms that is characteristic of the hysterical personality, although the symptoms themselves are triggered by foods and chemicals.[81] One early paper on hysteria indicated that 48 percent of hysterics reported food intolerances.[79,82] Data from controlled food challenge tests would help determine the proportion of hysterics who have chemical- or food-induced symptoms.

Another recent observation is that environmentally-ill patients may exhibit signs and symptoms of acute organic brain syndromes that resemble intoxication. For example, a typical patient with central nervous system symptoms may report difficulty in concentration and may exhibit confusion, poor short-term memory, slurred speech, motor incoordination, irritability, emotional lability, and flushing. Some epidemiologic studies on painters with chronic organic solvent exposure have shown significant cognitive impairments.[11,15,83] Controlled neuropsychological investigations of acute and chronic food and chemical reactions are needed to document the presence or absence of cognitive deficits in clinical ecology patients.

Clinical ecologists have noted that food, chemical, and inhalant factors are also related to certain somatic illnesses in addition to CNS and psychiatric syndromes. These include the nephrotic syndrome,[47] post-streptococcal glomerulonephritis,[84-86] cardiac arrhythmias,[43] thrombophlebitis,[45] and vasculitis.[44] Another report has noted abnormal incidence of primary hypothyroidism in workers exposed to polybrominated biphenyls in the manufacture of fire retardants.[87] Other somatic disorders which reportedly respond to clinical ecology treatment include rheumatoid arthritis,[2,88] migraine headache,[89,90] and ulcerative colitis.[52,91] In these latter illnesses medical thinking has previously emphasized psychological rather than chemical stress factors.[92]

Case Histories

Two case histories are given to illustrate symptom patterns and response to CE treatment. The first is the case of a nine-year-old boy with a complex history including rhinitis, periorbital edema, intermittent eye pruritus, urinary urgency and frequency, nocturnal cough, irritability, hyperactivity, and crying without provocation. The boy craved dairy products, and out-patient diagnostic tests suggested that milk and food colorings could trigger his symptoms. Double-blind tests, using both challenge feedings with capsules of milk powder and sublingual neutralizing dilutions of milk and placebos, were consistent with the clinical observations. Remission of symptoms occurred on avoidance of milk and food colorings.[93]

Another case is that of a fifty-one year old woman with history of cardiac arrhythmias, visual difficulties, recurrent bruising, cognitive dysfunction, and weight gain. The patient noted arrhythmias on exposure to chemicals in her job in an etching room of a factory. During challenge tests in an environmental hospital unit, numerous foods and chemicals (e.g., perfume, phenol, cigarette smoke, formaldehyde, insecticide) were found to trigger symptoms. She is reportedly symptom-free while following a rotation diet of safe foods and living in a non-urban area, but relapses on return to the city.[94]

CHAPTER FOUR

Possible Mechanisms in Environmental Illness

Patients with ecologic illnesses have syndromes that many traditional clinicians would consider "stress-related."[95] The stresses of poorly-tolerated environmental substances may combine with emotional stresses to lead to illness. Researchers on psychosomatic and psychiatric illness have begun to explore the possible biological mechanisms by which mind and body interact to produce disease. Figure 1 shows a schematic model of environmentally-induced illness. The hypothesis is that diverse external stimuli—such as natural inhalants, foods, chemicals, physical factors, and psychosocial stresses—can converge on some common pathways in the body early in the genesis of disease. The later divergence in the expression of particular outcomes or illnesses within the population depends upon the inherent tendency of the individual patients toward susceptibility in particular target organ systems. A patient with a family history of manic-depressive illness may show a bipolar affective disorder when he or she becomes sensitive to foods and chemicals, whereas another with a family history of autoimmune diseases might develop rheumatoid arthritis with his or her specific food and chemical sensitivities. The offending foods and chemicals might even be the same for these individuals with different diagnoses.

Two systems in the body that are obvious interfaces between external environment and internal milieu and which may both be involved in environmental illness are the immune system and the central nervous system.[96,97] Available data suggest that various environmental factors can impinge on both of these systems, which may then serve as part of the common pathway through which external stress promotes disease.

Immune System Mechanisms

Some workers have commented that the symptoms and syndromes of environmental illness could reflect mobilization of inflammatory pathways.[46,63] This mobilization could have both immune and nonimmune triggering mechanisms.[63,98] Regardless of how the events of inflammation are initiated in the earlier steps of the disease process, the end-result tissue dysfunctions would nevertheless look the same. In the past, interest has focused on nonimmune triggering factors.[63,99,100] Recently, more attention has been directed at immune system abnormalities in environmental illness.

Studies indicate that certain environmentally-ill patients have abnormalities in their immune system.[43-46,101-108] These are reflected in abnormalities of the functions and/or numbers of T-cells (the cellular branch of the immune system). Because T-cells regulate the humoral branch of the immune system (B-cell derived antibodies), it is not surprising that abnormalities in antibody and complement levels have also been noted.

No single abnormality appears in all patients, but certain trends have been reported. First, a number of observers have reported that in the majority of environmentally-ill patients, serum IgE levels are normal or even low,[101,105,109,110] although some (10-30 percent) reportedly do have elevated IgE levels. IgE is the main immunoglobulin that has been linked with Gell-Coombs Type I anaphylactic, immediate reactions, i.e., reaginic or classic allergy. Thus the majority of chronic food and chemical sensitivities treated by clinical ecologists are probably not *reaginic allergy* in the strictest immunological sense.

Other abnormalities of immune function, however, are present in certain CE patients. For example, some workers have noted higher titers of immune complexes in food- and chemical-sensitive patients than in normals.[102,103,107] One group has identified IgG type immune complexes in their most severely ill patients during acute exacerbations.[102,103] Immune complexes are also found in autoimmune disorders such as systemic lupus erythematosus and rheumatoid arthritis.

Recent studies have found immune complexes and early rheumatoid-like joint lesions in rabbits fed cow's milk or injected with bovine serum.[111-114] Furthermore, certain monkeys fed alfalfa sprouts or L-canavanine sulfate, a nonprotein amino acid constituent of alfalfa sprouts, reportedly develop a syndrome similar to systemic lupus erythematosus with corresponding immune abnormalities.[115] In both rabbit and monkey studies, there were individual differences in susceptibility to adverse reactions from ingestion of these foods.[111,115]

Immune complex disease likewise has been reported in workers exposed to vinyl chloride.[116] Some investigators have also noted an association between history of organic solvent exposure and acute poststreptococcal glomerulonephritis, an immune complex-related disease.[84-86] These findings indicate that hyperimmune phenomena, involving Gell-Coombs Type III reactions, may occur in some patients with environmental illnesses.[107]

It is significant that rats with experimentally-induced immune complex disease have been found to have immune complex deposits in the choroid plexus of the brain, with concomitant behavioral changes.[117,118] These data may provide clues to the etiology of the psychiatric aberrations seen clinically in some patients with systemic lupus erythematosus and other autoimmune disorders, as well as, perhaps, with environmental illness.

Some clinicians have also reported an unusually high prevalence of low serum IgA among patients with food and/or chemical sensitivities.[110,119,120] IgA is the immunoglobulin which plays a protective role in the gut, liver, and mucous membranes against the entrance of foreign substances into the body. A low IgA level presumably would permit easier access to the body for antigens and other such substances, which could then stimulate an increased response in other parts of the immune system.[120]

Immune complexes as well as nonimmunologic factors can activate the complement system. Complement is a set of serum proteins that interact to produce elements of the inflammatory response as well as destruction of foreign cells, bacteria, and viruses in the body. In patients with environmental illness, Rea has found evidence of abnormal activation of the complement system, including changes in the levels of total hemolytic complement,[43,46] C3,[44,45] and C4.[45] Other workers have reported similar data, which they consider suggestive of Gell-Coombs Type II, cytotoxic reactions.[105] Complement function is one of the potential common pathways on which antigenic, chemical, physical, and psychosocial stressors all might act, leading to the same set of endogenous events which cause the symptoms of environmental illness.[63,99,100,121]

Some writers have hypothesized that imbalances in the function of the cellular branch of the immune system may leave the patient open to a wide variety of disorders, including infections and autoimmune diseases.[96,97] T-lymphocytes, which are responsible for cellular immunity and regulation of other inflammatory cells, may also exhibit a variety of alterations as a result of adverse reactions to foods or chemicals, as well as to psychosocial stresses. Many of the most severely ill CE patients have low absolute T-cell counts,[43,46] both chronically at baseline and acutely with challenge test reactions. Some clinical ecologists suggest that the suppressor T-cell population is particularly affected, but data on this point are inadequate. Suppressor T-cell dysfunction has been noted in migraine patients,[122] asthmatic children,[123] and adults with seasonal rhinitis.[124] These studies did not address the possible role of foods and chemicals in the illnesses of the individuals under study. Consequently, there is no direct evidence to date of suppressor T-cell dysfunction in food- and chemical-sensitive patients with diagnoses such as migraine or asthma.

Recently, the release of a T-cell lymphokine, LIF (leukocyte inhibition factor), has been demonstrated in patients with nonreaginic (i.e., non-IgE-mediated) adverse food reactions to milk or corn.[104,110] (Lymphokines are soluble substances, released by antigen-sensitized lymphocytes, which regulate the activity of other white blood cells that participate in inflammatory reactions.) In another study, patients with chronic urticaria (hives) due to specific foods and food additives showed *in vitro* evidence that their T-cells were mobilized by offending agents but not by control substances.[108] One report on children with the allergic tension-fatigue syndrome noted *in vitro* positive lymphocyte transformation tests with specific foods and food additives. They also observed clinical remission on avoidance of the positive items in the majority of patients.[125] Since no blinding of the clinical trials and no control groups for the laboratory or clinical measures were employed, the usefulness of these data is limited. Taken together, nonetheless, these data suggest a possible role of Gell-Coombs Type IV delayed hypersensitivity reactions, mediated by the activation of T-cells, in chronic ecologic reactions.

Experimental evidence is available indicating that T-cell function may be compromised by certain stressors. For example, one acute exposure to 0.4 ppm ozone, an amount equivalent to a very smoggy day in Los Angeles, can impair the ability of T-cells in normal young male subjects to respond to a standard *in vitro* test for mitogen activation of those cells.[126] Recovery of normal function may take as long as two weeks after the single ozone exposure. Other environmental chemicals from foods, food additives, and contaminants, may also alter cellular and humoral immune system activity by direct immunotoxic effects.[127,128] Such chemicals include gallic acid from the food additives propyl gallate and tannic acid (e.g., from tea, coffee, cocoa);[129,130] BHA (butylated hydroxyanisole, a common antioxidant food additive);[131] carrageenan;[132] and certain pesticides.[133,134] The question remains unresolved as to whether reactions to these chemicals are due to non-specific irritant or toxic effects or due to immune sensitivities.

There is also growing evidence that psychosocial stress is sometimes correlated with impaired T-cell function.[96,97,135] For example, impaired mitogen responsivity of T-cells has been found in recently bereaved spouses.[136] T-lymphocytes may therefore be another common pathway in the body by which chemical, antigenic and psychosocial stressors can all affect the activity of the immune system. The full implications of these findings for human susceptibility to various illnesses are unresolved at present, however.

A common finding is that a CE patient may have symptoms in multiple areas of the body. Transient pains, inflammation, edema, smooth muscle dysfunction (such as blood vessel spasms), heart arrhythmias, and CNS disturbances may all appear in the same individual.[43] The multiplicity of symptoms raises a question: How can so many different abnormalities become manifest in reaction to a single substance? One hypothesis is that certain endogenous mediator substances, activated by both immune and nonimmune mechanisms,[98] trigger local dysfunction in their respective target tissues all over the body. Alternatively, adverse reactions to environmental substances could involve abnormally increased sensitivity of end-organ receptors to specific mediators *per se*.

A number of observers have found direct and indirect evidence of abnormal release or action of mediators in patients with adverse food or chemical reactions—e.g., prostaglandins,[101,137] kinins,[63,99,138] serotonin,[103,139-141] histamine,[142,143] and acetylcholine.[144] Other studies suggest that several of these mediators may interact by stimulating each other's release, potentiating one another, or counteracting each other's actions.[63,143,145] Moreover, preliminary reports suggest that T-cells have receptors for endogenous opiate-like mediators such as enkephalins[146] and that enkephalins may modify T-cell function.[147]

These latter data suggest a physiological means by which psychosocial stress could affect cellular immunity in triggering enkephalin release by the adrenals.[147]

These findings suggest that ecologic reactions may involve a network of different pathways associated with various endogenous mediators. The mediators can trigger symptoms in susceptible areas of the patient's body. The vulnerability of these areas would depend on such factors as inborn predispositions,[4,5,148] nutritional deficiencies,[149] prior local trauma, and prior viral and bacterial infections.

Central Nervous System Mechanisms

Research into the role of the central nervous system in abnormal responses to environmental substances has only recently begun. Already the data point to significant interactions between the immune system and the central nervous system.[96,97] Lesions or electrical stimulation of certain areas of the hypothalamus can alter both cellular and humoral immunity.[150-153] These effects include decreased susceptibility of experimental animals to death by Type I anaphylactic shock triggered by antigens,[151] or by mediators such as histamine[152] or bradykinin.[153] Another study has shown changes in neuronal firing patterns of cells in the ventromedial nucleus of the hypothalamus of rats in response to systemic administration of specific antigens.[154]

The hypothalamic lateral nucleus (LH) and ventromedial nucleus (VMH) have long been associated with control of eating behavior: LH is linked with increased eating and VMH with satiety. The VMH has also been linked to regulation of the morphine withdrawal syndrome,[155] and to insulin-induced hypoglycemic convulsions in animals.[156] These data imply that feedback loops between the immune and central nervous systems could be important for the maintenance of homeostasis when changes occur in the external or internal environment. In homeostasis, the hypothalamus helps regulate the autonomic nervous system, catecholamines (e.g., epinephrine and norepinephrine), and the endocrine system. In turn, the autonomic nervous system, catecholamines, and endocrine glands modulate the functions of many tissues that are affected in ecology patients, including various smooth muscles and the immune system.

At this point, however, it is premature to assume a direct cause-effect relationship between these observations and the clinical phenomena of environmental illness. One might speculate that the VMH in the hypothalamus regulates bodily awareness of various substances, including antigens, nutrients such as glucose, and drugs such as opiates, and thus the VMH may also mediate aspects of food and chemical addictions in CE patients.[63]

Despite the lack of conclusive biochemical data showing that food and chemical sensitivities can directly affect CNS function, one of the most significant reports of the clinical ecologists is that patients with environmental sensitivities often suffer from neuropsychiatric symptoms.[157]Although little evidence is available from studies on human subjects, animal experiments suggest numerous hypotheses for biological mechanisms. Some of these are:

1. Various foods supply neurotransmitter-precursors which affect brain levels of neurotransmitters, which in turn control certain brain functions and behaviors. Such precursors include tryptophan for serotonin; tyrosine for

dopamine, norepinephrine, and epinephrine; and choline for acetylcholine.[158]

2. Digestion of foods generates pharmacologically-active agents such as opiate-like peptides.[56] Some studies suggest that small amounts of opiate-like peptides can cross the blood-brain barrier.[159]

3. Entrance of food into the stomach and intestine releases digestive hormones and activates other hormones which can modify CNS neural activity and behavior.[56,59,60,63,160]

4. Food contaminants such as heavy metals and pesticides can exert direct toxic effects on CNS and peripheral nervous system.[161]

5. Immune complexes induced by food antigens may damage the cerebrospinal fluid/blood barrier and lead to behavioral abnormalities.[117,118] The mechanisms could involve immune damage to cells as well as activation of mediators affecting neuronal function. Some evidence also suggests that even some large protein molecules serving as antigens can cross the blood-brain barrier after gastrointestinal absorption.[162]

6. Particular foods in the digestive tract may stimulate specific neural signal patterns from gut to brain, especially the hypothalamus, and conversely, the response of brain to gut.[163,164]

7. The volatile chemicals that make up food odors can stimulate the olfactory system. A concept central to the understanding of food sensitivities is that foods are not only sources of nutrients, but also complex mixtures of organic chemicals. For instance, it is the unique pattern of chemical constituents that make a tomato a tomato rather than an apple.

Table 5 shows a partial list of natural chemical components of some common foods. Some of these chemicals have their own specific effects on the body. Apart from their role as nutrients, foods and their metabolites may act as foreign antigens; as pharmacologically-active chemicals in the body; or as olfactory stimulants for the central nervous system.

The olfactory system has known neuroanatomical and neurophysiological links to the hypothalamus, especially the lateral hypothalamic nucleus, and other areas of the limbic system.[165-168] This author has previously speculated that the olfactory system, hypothalamus, and limbic system pathways would provide the neural circuitry by which adverse food and chemical reactions could trigger certain neuropsychological and psychiatric abnormalities.[169]

Clinical ecologists have reported anecdotally that many patients begin food binges, violent behavior, or hypersexual activity after breathing non-food chemicals such as diesel fuel, organic solvents, and even detergents to which they are sensitive.[69,119,169] Other clinical observations include changes in apparent olfactory thresholds of environmentally-ill patients during diagnostic and therapeutic manipulations of food and chemical exposures. The sense of smell often becomes more normal (from hypo- or hyper-acute) as treatment progresses. Some neurophysiologists have noted that these olfactory-limbic system pathways provide a means by which food or chemical odors could affect many behaviors, including eating, drinking, reproductive behavior, motor activity, mood, cognition, and memory.[165,168] The limbic system is known to play a role in the regulation of these functions. However, the clinical implications of the possible links between behavior and the olfactory system in human subjects are largely unexplored.

Finally, another important hypothesis for a mechanism of environmental illness is that of classical conditioning. The patient's symptoms might be, in part, a conditioned response to specific sensory stimuli, including olfactory ones.[170] The initial symptoms could have been physiologically-based, unconditioned reactions to a given environmental exposure. Subsequently the patient might begin to experience similar symptoms in association with conditioned stimuli that otherwise would not physiologically induce such symptoms. The sight of a pesticide can, for example, or the faintest odor of natural gas might elicit a severe exacerbation of symptoms by conditioning rather than by strictly

Table 5. Partial List of Chemical Constituents of Tomato, Apple, Milk, and Orange

Tomato†

Component	*Olfactory Threshold (parts per billion)*
hex-cis-3-enal	0.25
deca-trans, trans-2, 4-dienal	0.07
dimethylsulfide	0.33
B-ionone	0.007
linalool	6
guaiacol	3
methyl salicylate	40
2-isobutyl thiazole	3.5

Milk*

p-cresol
4-ethylphenol
3-n-propylphenol
phenylacetic acid
hippuric acid
caprylic acid
palmitic acid

Apple‡

Component	*Olfactory Threshold (parts per million)*
ethanol	100
hexanol	0.5
hexanal	0.005
2-hexanal	0.017
butyl acetate	0.066
2-methylbutylacetate	0.005
ethyl 2-methylbutyrate	0.0001
hexyl acetate	0.002

Orange°

α-pinene
myrcene
limonene
linalool
cis-2, 8-p-menthadien-1-ol
decanal
carvone
valencene

†Buttery, R.G., Seifert, R.M., Guadagni, D.G., and Ling, L.C. Characterization of additional volatile components of tomato. *J Agr Food Chem.* 19(3):524-29, 1971.

‡Flath, R.A., Black, D.R., Guadagni, D.G., McFadden, W.H., Schultz, T.H. Identification and organoleptic evaluation of compounds in Delicious apple essence. *J Agr Food Chem. 15(1):29-35, 1967.*

*Brewington, C.R., Parks, O.W., Schwartz, D.P. Conjugated compounds in cow's milk-II. *J Agr Food Chem.* 22:293-94, 1974.

°Moshonas, M.G. and Shaw, P.E. Composition of essence oil from overripe oranges. *J Agr Food Chem.* 27(6):1337-39, 1979.

biological effects. Such conditioning phenomena, dependent on the familiarity or unfamiliarity of environmental cues, have been reported even with powerful, pharmacologically-active agents such as morphine or heroin.[171,172] Classical conditioning of immune system suppression has also been reported.[96,173] The possible role of classical conditioning in adverse ecologic reactions remains unstudied to date.

CHAPTER FIVE

Diagnosis of Environmental Illness

FOR THE PATIENT IN WHOM environmentally-induced illness is a possibility, the clinical approach begins with a traditional medical and psychiatric history, physical examination, and appropriate routine laboratory tests, including white blood cell count and sedimentation rate. Additional laboratory studies often include levels of absolute T-cells, immunoglobulins, total complement as well as C3 and C4, and immune complexes of the IgG type. Mediators such as histamine, serotonin, and prostaglandins are also, at times, evaluated.

Symptoms of the allergic toxemia syndrome such as fatigue, difficulty in concentration, or musculoskeletal complaints suggest looking more closely for environmental factors in the patient's problems. Some adult patients who present with psychological or systemic symptoms may also give a history of having "outgrown" childhood allergies with physical symptoms such as hives, rhinitis, or asthma. Physical findings are rarely unique to environmental illness, although signs such as facial pallor, enlarged cervical lymph nodes, and "allergic shiners"—dark, purple, edematous tissue below the eyes, without evidence of infection—are a common aspect of environmental illness, especially in children.[174]

Dietary and Environmental History

The history includes the patient's dietary habits and environmental exposures in extensive detail. Patients generally fill out lengthy questionnaires comprising checklists of commonly eaten foods as well as of indoor and outdoor chemicals in home, hobby, school, and work environments (see Appendix for sample questionnaire). Exposure history is, at best, only a rough guide to the initial choices of items for which a patient should be tested. Other significant clues to the role environmental substances play in the patient's problems include the following:

1. History of a strong craving for or strong aversion to a particular item.
2. History of temporary relief of symptoms when a patient is exposed to an item after a period of avoidance of less than four days.
3. History of an increase in symptoms after unaccustomed overexposure to an item.
4. History of frequent exposures to an item (more often than once in four days) or inordinately high quantities per exposure.

All of these may justify testing for a sensitivity, regardless of whether or not the patient is aware of any effect of the item upon symptoms. Chronic adaptation may have attenuated any acute symptom flares.

High risk foods are those that are common and often are included as hidden ingredients of commercial products, e.g., corn, yeast, wheat, milk, and egg.[39] In addition to food additives and outdoor air pollution, high risk chemicals include fumes from natural gas stoves and heating systems;[175-179] tap water contaminants;[180] indoor pesticide powders and sprays; outgassing from soft plastics in home and office furnishings and food storage containers;[181,182] and formaldehyde in permanent-press fabrics and particle board[183-185] (see Table 2).

Other important aspects of the history include noting any changes in symptoms with changes in seasons or with vacations taken away from customary home environments. Such changes in the environment may lead to alterations not only in psychological, but also in chemical and antigenic exposure factors. Some patients can trace the onset of illness to the aftermath of a severe influenza infection; a massive exposure to a particular toxic chemical, as might occur in an industrial accident;[46] an exposure to improperly-ventilated home-use chemical products; a move to a new home; or a severe psychosocial stress.

In certain severe cases, a "spreading phenomenon" develops; the number of offending items grows with alarming and seemingly relentless rapidity until it peaks and stabilizes. A sensitization to one massive chemical overexposure often triggers this spread of sensitivities. At that point, minute exposures to many other chemicals all begin to set off the multiple-system symptoms in these patients.

Clinical Tests

Once the history has suggested possible areas for evaluation, clinical tests for specific food, chemical, and natural inhalant sensitivities can begin. Tests start with the most highly suspect items, then proceed to the more subtle aspects of the individual's total load of environmental contacts. Clinical ecologists differ widely among themselves in their practices; they emphasize different aspects of the diagnostic and therapeutic approaches discussed below.

The major diagnostic techniques fall into three general categories: (1) food or chemical avoidance, usually followed by a challenge test; (2) serial dilution titration or provocation-neutralization tests using intradermal injection or sublingual drops of diluted extracts of test items; and (3) *in vitro* cytotoxic blood tests.

Avoidance followed by challenge tests

Some clinicians prefer to diagnose food sensitivities simply by having the patient eliminate a particular food from the diet for several weeks and noting any improvement in symptoms.[52,186-188] Individuals with limited financial resources for obtaining medical care, as well as those with less severe health problems, often find elimination diets to be extremely valuable. They offer a simple and rapid means of establishing the presence or absence of a food sensitivity problem and may even be adequate for treatment in less severe cases.

Clinical ecologists, however, generally prefer specific challenge tests.[39] In the challenge test, the patient avoids the suspected item for a period of 4-5 days and then undergoes acute re-exposure to a pure form of that single food or chemical. Since avoidance for more than a few days may improve tolerance, a challenge test done after 12-14 days may fail to elicit symptoms. If the patient's sensitivity to the item remains even after this period of avoidance, he or she will experience an acute attack of characteristic symptoms. Such attacks usually begin within 1-2 hours after ingestion of the offending food, although onset may be delayed up to 24 hours, and symptoms may last from several hours to several days. Some patients reportedly exhibit increases in heart rate or heart rate variability after exposure to an offending substance.[43,71,119,189-191]

The access routes and rate of absorption, as well as the nature of the test substance can affect the speed of onset of symptoms and the nature of a reaction. For example, meals with high fat content will be absorbed more slowly from the gut and hence result in slower onset of symptoms than low-fat meals. Randolph[1,9,66,69] believes that chemicals are generally more potent than foods, which in turn are more potent than natural inhalants in triggering severe symptoms—especially those in the CNS. He also reports that inhaled chemicals can set off reactions very rapidly, often within seconds or minutes after exposure.

A major advantage of the challenge test is that it can be very convincing to the patient. A symptom flare caused by an offending substance can provide positive evidence for those who otherwise might resist making the drastic changes in diet and chemical environment necessary for effective treatment. Challenges can be repeated periodically to gauge the return of tolerance. Such tests are inexpensive to perform, requiring no special testing materials other than a sample of the suspected food or chemical. Patients can often test chemicals by sniffing briefly from a sample kept in a closed jar. Challenge tests involve patients actively in their own diagnostic evaluation and teach them skills that they can later use in self-care and self-reassessment.[187]

A number of lay books present the principles of clinical ecology, including challenge tests, in excellent detail (see Resources section). CE is unusual in the medical world for its emphasis on the active part that patients must play in the successful evaluation and management of their own illnesses, using physicians mainly as guides.

Challenge tests have several disadvantages, however. In the first place, a maximum of three or four tests, separated from one another by hours, can be done in a day, and only one or two can be done if some items are poorly tolerated. Challenge testing also involves stress on the patient, who may experience intense symptom flares. In addition, it is quite difficult to blind the challenge tests so that neither the patient nor the person administering the test knows the identity of the test item. This problem has been solved in the research setting with the use of nasogastric tubes or gelatin capsules containing the test food.[48,192] Clinical tests are generally not blinded. Consequently it is difficult to separate the effects of subjective expectation—on the part of both the patient and the physician—from the biological effects of a test substance.[193-195]

Serial Dilution Titration

One of the most controversial aspects of CE is the testing and treatment techniques that employ dilutions of antigenic extracts, i.e., serial dilution titration and provocation neutralization methods. The serial dilution titration method uses skin tests both to identify substances to which an individual is sensitive and to determine the dilution of antigen with which to begin treatment for adverse reactions to natural inhalants such as pollens, dusts, and molds. The technique was first developed in the 1940's by the allergist Herbert Rinkel[196] as a modification of an approach begun by otolaryngologist-allergist French Hansel.[197] It uses the simultaneous intradermal application of a set of progressively stronger 1:5 dilutions (e.g., 1:78,125, 1:15,625, 1:3125, 1:625, etc.) of the test material concentrate, usually in doses of 0.01 cc.

In the usual case, all test doses initially form a 4 mm skin wheal. If the patient is sensitive to the test substance, the wheals with the stronger dilutions will grow over a 10 minute period. The dilution that first induces a wheal growth of 2 mm is called the endpoint. Treatment begins with some volume of that dilution. A typical test pattern would exhibit a sequence of wheals measuring 5 mm, 5 mm, 7 mm, and 9 mm ten minutes after the injection of their respective dilutions (see Figure 3). The endpoint in this case would be the dilution associated with the 7 mm wheal, which would often be surrounded by erythema. In serial dilution titration the skin whealing response alone—not a change in symptoms—is the criterion for selection of the appropriate treatment dilution.

Provocation-Neutralization Tests

Provocation-neutralization (P-N) testing was originated by allergists Carlton Lee[198] and Herbert Rinkel,[199] and further developed by various clinical ecologists.[2,200] It is the most controversial aspect of clinical ecology and has led to heated debate.

Provocation-neutralization testing is a technique which diagnoses a sensitivity by assessing the ability of a test dose to evoke symptoms rather than to induce wheal growth alone. Like the Rinkel method for inhalants, the P-N test utilizes serial dilutions, in a 1:5 ratio, of each substance. The usual P-N test dose, however, is larger—0.05 cc—and is administered either intradermally or sublingually. In contrast to the serial dilution titration method, P-N testing employs

Figure 3. Skin Whealing Response in Serial Dilution Titration

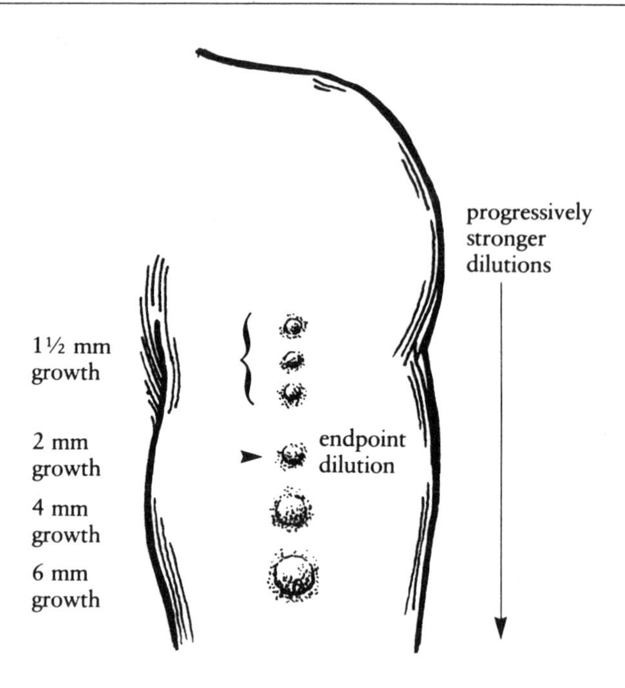

In serial dilution titration a series of progressively stronger dilutions of extracts of a suspected offending substance is injected simultaneously. All the dilutions produce wheals which are originally about the same size (4 mm). During the 10 minutes following injection, the wheals grow in diameter; the weaker dilutions produce little growth while the stronger dilutions produce progressively more growth. The first dilution (in the progression from weak to strong) that produces a wheal growth of 2 mm is called the endpoint dilution. The endpoint dilution is used to determine initial treatment dosage.

only one dilution of a single substance at a time. The P-N method also applies to a broader range of test materials—including not only pollens, dusts, and molds, but also foods, chemicals, drugs, hormones, and even viral vaccines.[2,200]

The goals of this method are first to identify substances that provoke symptoms in an individual, and second to discover which dilutions of those substances are appropriate to use in treatment. Empirically, clinical ecologists have found that certain dilutions of an offending substance will provoke and other dilutions will relieve mild versions of the patient's characteristic symptoms during a ten minute test period.[200] In this method, the neutralizing dose (ND) is a symptom-relieving dilution. *Both* stronger (overdose) and weaker (underdose) dilutions can reportedly provoke symptoms. Thus, the dose-response curve in P-N testing is often non-monotonic, that is, biphasic or even sinusoidal rather than linear (see Figure 4).[200] Significant precedents for the biological phenomenon of non-monotonic dose-response curves are documented, but remain unacknowledged, in the mainstream research literature.[68, 201-205]

An illustration of how P-N testing works is as follows: A patient presents with a history of chronic migraine headaches, fatigue, depression, and episodic abdominal cramps. This individual gives a history of daily consumption of eggs over a period of many years. During P-N tests, seven minutes after the 1:5 dilution (#1) of whole egg extract is injected intradermally, the patient complains of a new onset, mild left-sided headache, sadness, and fatigue, with some abdominal discomfort. These symptoms persist for ten minutes after the test dose. The 1:25 dilution (#2) is then administered, and the patient reports a gradual clearing of all these symptoms over the next ten minutes. The #1 dilution would be considered an overdose and the #2 dilution a neutralizing dose.

When a 1:125 dilution (#3) is given, the patient reports an abrupt onset of faintness, a strong left-sided headache, severe abdominal cramps, and crying within two minutes after the injection. This would be considered an underdose response. As a result, the patient is given another injection of the #2 dilution (neutralizing dose); and, in the ideal case, the faintness, headache, cramping, and

Figure 4. Sample Nonmonotonic Dose-Response Curve in P-N Testing

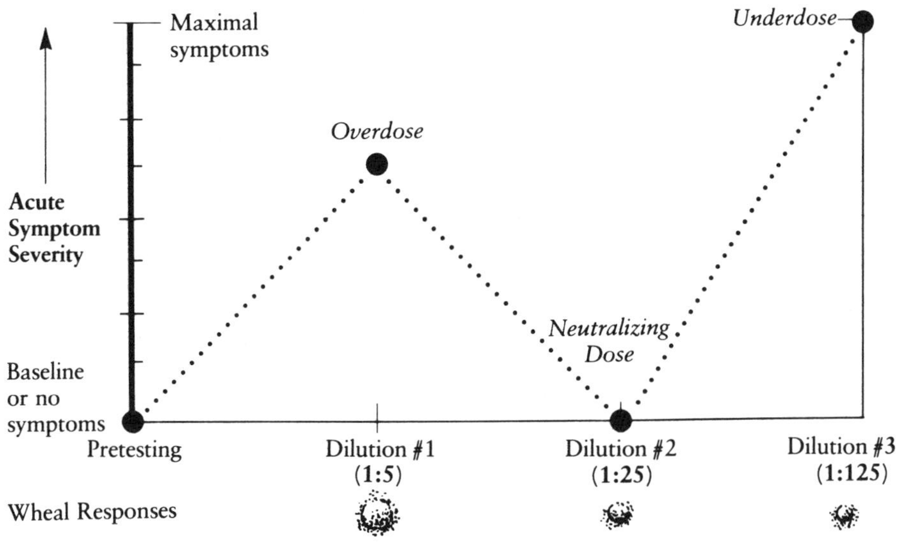

In this case, both dilutions #1 and #3 provoked symptoms while dilution #2, intermediate in concentration between the latter two dilutions, relieved or "neutralized" symptoms. Furthermore, the severity of induced symptoms was actually greater with the less concentrated (1:125) than with the more concentrated (1:5) dilution. However, wheal size and hardness progressively decreased with decreasing concentration of the test dose.

tearfulness would cease within ten minutes. It is notable that the severity of symptoms reportedly is greater and speed of onset of acute reaction sometimes faster after a weaker than after a stronger provoking dose.

Intradermal tests with 0.05 cc doses begin with a 7 mm wheal on injection. At the ten-minute point after injection, clinical ecologists report that each wheal has a characteristic contour, color, hardness, and increase in size. The wheals for overdose, neutralizing dose, and underdose each exhibit a certain set of differential characteristics that correlate with the symptom responses (see Table 6). The reader is referred to publications on this method for details of wheal patterns.[2,198-200]

Some clinical ecologists use wheal response alone or in combination with symptom reports to determine the neutralizing dose. Others employ sublingual drops of the test dilutions instead of injections. With this approach, symptom provocation and neutralization are the only criteria for diagnosing sensitivities. Children and other patients who are unable to tolerate repeated injections are given sublingual rather than intradermal tests. Food colorings and certain other chemicals are routinely tested by the sublingual method.*

The test materials are purchased from commercial allergy supply houses. They are mixed with standard chemical preservatives such as phenol or glycerine to prevent bacterial contamination and maintain antigen stability. Some patients reportedly are sensitive to the preservatives and require either neutralization for those chemicals or even preservative-free extracts. Without the preservatives, however, even if the extracts are refrigerated or frozen between uses, there is a risk of bacterial contamination and loss of biologic or antigenic activity in inhalants and foods. Systematic studies on the stability and purity of preservative-free extracts are needed for determining the validity of using them in the treatment of the chemically-sensitive patient.

Table 6. Characteristics of Typical Wheals

Overdose	*Underdose*	*Neutralizing Dose*
Growth of 2 mm or more	No or little growth	Growth up to 1½ mm
Markedly raised	Flat	Flat or slightly raised
Markedly hard	Soft	Soft to medium
Blanched	Neutral color or pink	Neutral color or pink
Thick discoid form, with sharply demarcated cliff-like edges	Smoothly tapering edges	Irregular, ragged, frayed edges

SOURCE: Miller, J.B. *Food Allergy: Provocative Testing and Injection Therapy.* Springfield, Illinois: Charles C Thomas, Publisher, 1972, p. 44.

*Interestingly, substances such as radioactively-labeled glucose, pesticide, and sodium chloride have been demonstrated to travel from the ligated oropharynx to the intracranial cavity of rats within five minutes. This route from mouth to brain bypasses the gut and leads to higher brain concentrations than from direct gastrointestinal administration of the same substances.[206] The relevance of these findings to clinical sublingual P-N tests is unknown, but merits investigation.

The data on the reliability and validity of the P-N testing technique are still inconclusive at this time. While both positive[71,207-209] and negative findings[210-214] have appeared in the literature, major methodological and interpretive problems in existing negative studies make it difficult to draw firm conclusions.[215,216] Furthermore, most of these studies have tested the efficacy of the overdose provocation but not of the underdose provocation or of the neutralization technique.

The potential advantages of the P-N method make it an important area for further scientific study. In P-N testing, the patient can usually continue to eat his or her customary diet, live in the usual chemical environment during the testing, and endure less severe reactions than with challenge tests. There is also less potential for results to be affected by suggestion. The P-N approach is more amenable to single-blinding in the clinical setting and to double-blinding for experimental purposes than are challenge tests.[71] The disadvantages, however, include the lesser participation of the patient in the diagnostic process; the significant cost of office testing of a large number of items; and the unresolved questions about whether or not P-N can reliably reproduce the results of a challenge test.

Cytotoxic tests

The cytotoxic blood test is based on the principle that extracts of foods and chemicals to which the patient is clinically sensitive will induce visible damage to white blood cells. Technicians can grade the damage on a 0 to 4 + scale of severity, on microscopic examination. The Bryans[217] first developed the cytotoxic test for widespread clinical use from Black's original concept.[218] Some investigators have used it in screening blood of large numbers of patients, without the difficulties of the more tedious item-by-item *in vivo* testing methods.[219,220]

Other researchers have demonstrated problems with the validity and reproducibility of the test results.[221] One criticism is that this technique is susceptible to subjective judgments of the technicians and to variations in methods used by different laboratories. Recent studies, however, have found evidence that the method is indeed reliable on test-retest.[222,223] On the other hand, evidence for validity—i.e., a strong correlation between results of cytotoxic and the corresponding challenge tests with specific foods—does not yet exist.[222] A 4 + result on cytotoxic testing, for example, may be associated with only mild clinical symptoms (e.g., a few hives), whereas a 1 + cytotoxic test result might be linked with a severe set of symptoms (e.g., severe asthma) on actual ingestion.

Most clinicians who perform cytotoxic tests use only one blood sample from a given patient and make long-term recommendations for avoidance of items that test positive. The clinical usefulness of a single cytotoxic test is unestablished. The point in time for retesting is also unclear. Treatment that follows from a cytotoxic test is limited to avoidance; it gives no indication of when or if the item would be tolerated at spaced intervals on a rotary diet schedule.

Reliable and valid *in vitro* tests for non-IgE-mediated food and chemical sensitivity would be a valuable part of the diagnostic work-up, but no such test is available at present. Standard RAST tests (radioallergosorbent) used in traditional allergy work are helpful only in patients with IgE-mediated allergies to the test materials. Another *in vitro* assay, based on red blood cell lysis and possibly mediated by IgG or IgM, is in the developmental stage.[224,225] Additional tests may emerge from the reported changes in T-cell function using food extracts *in vitro*,[108,110,125,126] but such assays are not ready for routine clinical application.

CHAPTER SIX

Treatment of Environmental Illness

THE PHILOSOPHY OF TREATMENT in clinical ecology is to avoid triggering symptoms in the first place rather than to suppress them with medications once they appear. As a result, the main methods of ecologic therapy include: (1) avoidance of offending substances; (2) rotation diet of tolerated foods; and (3) neutralization or prevention of adverse reactions by subcutaneous injection or sublingual drops.

Avoidance

The patient eliminates from the diet and/or chemical environment the offending items which cannot be tolerated even infrequently. In practice, chemical avoidance is a major project, but often essential before other treatments can be effective. Costly alterations of patients' homes are sometimes required. It may be necessary to replace synthetic wall-to-wall carpeting with hardwood floors and all-cotton scatter rugs; substitute electric stoves for gas stoves; and replace oil and gas heating systems with electric heaters, radiant heating plates, or even solar heat. For the less ill patient, merely clearing a "safe haven," such as the person's bedroom, of all synthetic materials can provide a retreat for the patient during recovery from acute reactions. Many clinical ecologists feel that control of the chemical environment is the most important first step that patients must take before treatment for foods or other stressors is initiated, and that chemical avoidance alone may circumvent the need for additional measures.

Rotation Diet

This diet includes tolerated foods at a regularly-spaced interval of 4-7 days between exposures.[39,64] At first, rotary diets involve 3-4 meals per day, each consisting of a single food. Eventually, within 3-6 months, rotated multiple-food meals may be consumed (see Table 7). Foods that are taxonomically related to one another often cause similar reactions in patients with sensitivities (see Table 8). Hence, meals containing members of the same food family, such as wheat and rye, also initially must be separated by at least 3-4 day intervals.

The principle of rotation is that tolerated foods will not be eaten so frequently as to induce new sensitivities or increase any mild sensitivity to them. As tolerance returns for foods that have been avoided, those items may be worked into a rotation schedule without reactivation of sensitivity. In contrast, ordinary elimination diets involve eating a small number of tolerated items daily. Clinical ecologists believe that this frequent repetition can lead to the spread of sensitivity to previously-tolerated items, gradually leaving the patient with fewer and fewer foods to eat. Thus, the rotary diet can be more effective than the elimination diet because the former favors maintenance and expansion of the number of tolerated dietary items.

The possibility of liberalizing the diet over time compensates for the inconvenience of initially eating only simple, single food meals and the necessity of more planning and record-keeping than with an elimination diet. The proper execution of a rotary diet demands much effort and self-discipline on the part of the patient. For the most part, only the most severely ill individuals are motivated enough to adhere strictly to the diet for long periods. Their alternative is debilitating illness.

The effectiveness of a rotation diet in treatment has not yet been documented in experimental or epidemiological research. However, CE-oriented physicians report that years of widespread clinical application support its efficacy.[2]

Optimal Dose Treatment

This treatment derives from the Rinkel serial dilution titration method of testing.[2,196] Once the endpoint dilution is determined for the particular pollen, dust, dander, or mold antigen, then therapy begins with some fraction or multiple of volume of the 0.01 cc endpoint dose. The goal is to find a dose (e.g., five times the endpoint dose or 0.05 cc) which will induce symptom relief for as long as possible between injections, for example, up to 5-7 days.

Clinical ecologists report that the treatment which follows from the Rinkel serial dilution method permits coseasonal therapy (which begins during the season for a particular pollen or mold) and provides immediate relief of symp-

Table 7. Sample Rotary Diversified Diets

I. Single-Food Meals: Initial diet plan includes a set of tolerated or "safe" foods, which may be very limited in number at the start of rotary diet treatment. Quantities of each food may be large.

		Day 1	2	3	4
Meal	A	lamb	halibut	turkey	shrimp
	B	broccoli	avocado	spinach	carrots
	C	cantaloupe	pineapple	yams	raspberries
	D	cashews	filberts	sunflower seeds	walnuts

II. Multiple-Food Meals: Maintenance diet of multiple-food meals is possible within 3-6 months of starting rotation schedule as sensitivities to additional foods decrease with avoidance and those foods can be rotated without inducing symptoms.

Meal	Day	1	2	3	4
	A	apples rosehips tea tapioca	pineapple cashews comfrey tea	bananas walnuts ginseng tea	oranges Brazil nuts peppermint tea
	B	lettuce tomatoes green pepper mushrooms sunflower seeds sunflower oil vinegar	lentil beans brown rice carob honey	yogurt peaches	avocado celery carrots
	C	pork Swiss chard papaya	shrimp Chinese cabbage water chestnuts sesame oil lychee fruit	turkey squash okra honeydew melon	trout yams asparagus blueberries
	D	macadamia nuts	grapes	coconut	filberts

See also books in Resource section for details of designing rotation diets.

Table 8. Representative Food Families

Grass (Grain) Family
barley
  malt
  maltose
bamboo shoots
corn (mature)
  corn meal
  corn oil
  cornstarch
  corn sugar
  corn syrup
  hominy grits
  popcorn
lemon grass
  citronella
millet
oat
  oatmeal
rice
  rice flour
rye
sorghum grain
  syrup
sugar cane
  cane sugar
  molasses
  raw sugar
sweet corn
triticale
wheat
  bran
  bulgur
  flour
    gluten
    graham
    patent
    whole wheat
  wheat germ
wild rice

Cashew Family
cashew
mango
pistachio
poison ivy
poison oak
poison sumac

Legume Family
alfalfa (sprouts)
beans
  fava
  lima
  mung (sprouts)
  navy
  string (kidney)
black-eyed pea (cowpea)
carob
  carob syrup
chickpea (garbanzo)
fenugreek
gum acacia
gum tragacanth
jicama
kudzu
lentil
licorice
pea
peanut
  peanut oil
red clover
senna
soybean
  lecithin
  soy flour
  soy grits
  soy milk
  soy oil
tamarind
tonka bean
  coumarin

Grape Family
grape
  brandy
  champagne
  cream of tartar
  dried "currant"
  raisin
  wine
  wine vinegar
muscadine

Potato Family
eggplant
ground cherry
pepino (melon pear)
pepper (Capsicum)
  bell, sweet
  cayenne
  chili
  paprika
  pimiento
potato
tobacco
tomatillo
tomato
tree tomato

Pheasant Family
chicken
  eggs
peafowl
pheasant
quail

Bovine Family
beef cattle
  beef by-products
    gelatin
    oleomargarine
    rennin (rennet)
    sausage casings
    suet
  milk products
    butter
    cheese
    ice cream
    lactose
    spray dried milk
    yogurt
  veal
buffalo (bison)
goat (kid)
  cheese
  ice cream
  milk
sheep (domestic)
  lamb
  mutton
  Rocky Mountain Sheep

SOURCE: Golos, N., Golbitz, F.G., *If This is Tuesday, It Must Be Chicken or How to Rotate Your Food for Better Health.* Dallas: Human Ecology Research Foundation of Southwest (Webbs Chapel Rd., Dallas Tx. 75234), 1981.

toms using lower doses than conventional allergy treatment. This contrasts with the traditional allergy practice of preseasonal buildup to larger antigen doses, with the need for 1-2 cycles of such injections over 12-24 months before maximum symptomatic benefit begins. Although some negative studies of the Rinkel method have been published,[226,227] methodological criticisms (e.g., use of inadequate doses) have raised doubts about the validity of their conclusions.

Some clinical ecologists note the importance of cross-reactive, concomitant pollen and food exposures that occur with seasonal changes. Certain pollens reportedly cross-react with particular foods. For example, sensitivity to milk tends to be exacerbated during ragweed season and sensitivity to cereal grains during the grass season. Pollen and food sensitivities are treated simultaneously in order to reduce reactivity to each type of substance.

Neutralization Treatment

The provocation-neutralization testing described above not only detects sensitivities, but also determines treatment in the form of neutralizing or symptom-relieving dilutions of offending substances.[200] Treatment involves long-term subcutaneous or sublingual administration of neutralizing dilutions. The patient can self-administer the neutralizing doses when exposed to offending substances in order to prevent both chronic and acute symptoms. Some patients require the treatment before each meal; many others can maintain themselves with drops or shots once or twice a week. Highly sensitive individuals may need treatment for all foods in their diet, but the less severely ill reportedly do well with neutralizing doses for only the major inhalants, common foods, chemicals, and hormones (e.g., progesterone in women) in their total load.

Some clinical observers have suggested that certain factors compose such an important part of many patients' total loads that specific treatment for those factors may lessen the need to neutralize the patient for many other items. Such key substances reportedly include natural phenyl compounds in many foods,[228] and the fungal infectious agent Candida albicans.[229,230] Rigorous investigations of the phenyl compound and Candida hypotheses are not yet available.

Many patients, especially the chemically-sensitive, do not achieve optimal symptom relief from treatment drops or shots alone and must combine this approach with some degree of avoidance of offending substances or even a rotary diet. In fact, some patients have so few fully tolerated foods when they begin the treatment program that they must use a rotary diet of foods which cause the least severe symptoms, in combination with injections of neutralizing doses for those specific foods.

The advantage of the neutralizing treatment is that the patient can achieve symptom reduction or control with a program much less disruptive to his or her lifestyle than with the combination of rotary diet and chemical avoidance. The neutralizing dilutions for substances other than foods, such as chemicals, drugs, hormones, influenza virus, dusts, pollens, molds and animal danders, can reportedly also be used in short- or long-term therapy for symptoms triggered by these items.[2,200] Clearly, the neutralization technique, if further research confirms its efficacy, would offer an important means of treating patients for substances that they cannot or will not avoid. As with P-N testing, both positive [208,231] and negative[214] studies of aspects of optimal dose and neutralizing dose treatments have been published. Definitive work on the many claims for these therapeutic methods still has not been done.

The optimal dose and P-N methods differ a great deal from more traditional allergy techniques. In contrast to clinical ecologists, traditional allergists use injections of progressively increasing concentrations of inhalant allergens over a period of months to build up tolerance. They do not accept the idea that a particular dilution of an offending substance can relieve symptoms acutely and do so within minutes. Furthermore, the "build-up" method of treatment in traditional allergy is applied primarily to natural inhalants; this technique is considered ineffective in food allergy. Clinical ecologists, in contrast, utilize the P-N method in testing and treating sensitivities to several classes of substances, including inhalants, foods, chemicals, drugs, hormones, and even viral vaccines.[2]

In practice, clinical ecologists who employ provocation-neutralization methods or *in vitro* studies use them primarily for outpatient diagnosis and treatment. Under close medical supervision, many patients also safely carry out 4-5 day spring-water fasts at home to clear symptoms of persistent food sensitivities. Frequently, the most severely-ill CE patients start with outpatient diagnosis, but soon it becomes apparent that their problems are too extensive for outpatient management. Patients with severe, multiple chemical sensitivities may require hospitalization in specially designed environmental units.[66] There they begin testing and treatment with a strict chemical avoidance program and a single-food rotation diet. The simple elimination diets or neutralization treatment alone fail to bring their symptoms under adequate control.

Critics of the avoidance and rotation diet program maintain that patients can suffer long-term harm on two counts: (1) nutritional inadequacies can occur on the limited diet, and (2) recurrent hyper-acute exacerbations may be caused by incidental environmental exposures. In the first instance, dietitian analysis of multiple-food rotary diets suggests that nutrient intake can be adequate, [232] although theoretically a severely restricted rotary diet followed for many months could induce nutritional deficiencies.

The second criticism is perhaps more serious. According to clinical ecologists, abrupt isolation from a given environmental agent may "unmask" or accentuate an otherwise low-grade, chronic sensitivity. Re-exposure to the agent will then reportedly trigger an unusually severe attack of symptoms. Such "hyper-acute" susceptiblity is thought to fade over a period of roughly two to four weeks of avoidance of the offending substance.

In practice, however, some patients may emerge from treatment in an environmental unit in an unrelenting state of heightened acute sensitivity that may last for months or even years. These patients are often disabled by their symptoms and must live in environmental seclusion at remote mountain or seashore locales. Some clinical ecologists believe that certain of these patients may have unsuspected factors which perpetuate their hyper-susceptibility state. These factors include incomplete compliance with avoidance and rotation programs, chronic infectious states (e.g., Candidiasis), and various synthetic materials in implants or prostheses (e.g., for dental or orthopedic purposes).

Without systematic outcome studies on large, representative samples of environmental patients, it is difficult to judge the extent of or role of CE treatment in either malnutrition or long-term hyper-susceptibility problems. However, one follow-up report on a sample of thrombophlebitis patients indicated more sustained improvement in those on CE treatment than in those receiving conventional medical care.[233]

Acute adverse reactions to environmental substances reportedly can be shortened by a variety of methods (in addition to the neutralizing doses) that are relatively less toxic than many synthetic drugs:

1. Clinical ecologists have found that alkali salts (sodium to potassium bicarbonate in a 2:1 ratio) taken in water can attenuate reaction to tests or to unintentional exposures.[1,2,66] Alka-Seltzer Gold (which does not contain aspirin) is a readily available and more palatable form of these salts. Patients with renal or cardiac conditions should not use the alkali salts. When oral intake is impossible, inpatients can be given intravenous sodium bicarbonate.
2. Induced vomiting within the first hour after ingestion is useful in minimizing further digestion of a nontolerated food item. After the first hour, laxatives such as plain milk of magnesia can be helpful.
3. Removing the patient from the source of offending chemicals, particularly inhaled substances, is essential. Breathing 100% oxygen for several minutes can reportedly facilitate clearance of symptoms.[1]
4. Some clinicians also employ intravenous ascorbic acid and pyridoxine to shorten adverse reactions to foods and chemicals.[234,235]

Rigorous experimental study of the value of each of these treatment techniques in adverse environmental reactions is still needed. The first three methods above are nevertheless part of routine treatment in conventional emergency medicine for poisoning from certain toxic substances.[236]

At present, clinical ecologists have few tools for altering the patient's endogenous susceptibility to environmental substances. Some anecdotal reports recently have suggested benefit from experimental treatment with transfer factor, a low-molecular weight product of white blood cells which can stimulate immune responsivity in T-cells. Much more data from controlled trials is essential before extensive clinical use of transfer factor would be appropriate.

Some clinicians believe that nutritional supplements and digestive enzymes are helpful in lessening environmental sensitivities.[234] Clinical ecologists also feel that many patients such as certain "hypoglycemics" and schizophrenics who are being treated with megavitamins by their physicians, are actually ecologically ill and require CE diagnosis and therapy as well, for optimal improvement. Dietary supplements that are thought to be useful in prevention and treatment of food and chemical sensitivities include ascorbic acid, pyridoxine, niacin, vitamin E, trace minerals, and digestive (especially pancreatic) enzymes. Some physicians feel that environmentally-ill patients have inadequate digestion of foods and insufficient detoxification of chemical exposures by the liver. They believe that this permits abnormally large quantities of antigens[237,238] and toxins to be absorbed into the bloodstream, triggering adverse reactions. These practitioners emphasize nutritional and enzyme therapies to bolster the early steps of excluding and processing environmental substances.

Maximum purity in vitamin supplements is needed in order to minimize adverse reactions to food and chemical contaminants (i.e., colorings, flavorings, sugar, dextrose, cornstarch filler). Products for ecology patients are often prepared without fillers in clear gelatin capsules, which are often tolerated. In certain cases patients must remove the contents from the capsules. Even with the purest supplements, some individuals must rotate the vitamins at intervals of four or more days. Many CE patients cannot tolerate any vitamin supplements, especially early in treatment. Since susceptibility is sometimes reduced following a 3-6 month period of total avoidance, retesting of vitamins may be useful. More detailed discussion of the use of nutritional and enzyme supplements in environmental illnesses is found in other sources (see Resources section).

CONCLUSION

CLINICAL ECOLOGY POSES MANY CHALLENGES to the current assumptions and traditional practice of medicine. Clinical ecologists suggest that patients with marked susceptibility to their environment may be harbingers of the future; the ever-increasing number of synthetic chemicals in the environment may mean that increasing numbers of people will one day develop environmental illnesses. Efforts to increase the body's capacity for tolerating toxic environmental substances may prove successful in providing relief for the environmentally ill. Clinical ecologists believe, however, that it is crucial for modern societies to reduce significantly the total load of chemicals to which everyone is exposed. This lessening of total load might then permit more people to tolerate the natural chemicals that make up the food they need for nourishment.[239]

Critics of clinical ecology point out that most of the major hypotheses and clinical techniques in the field remain untested or unproven by rigorous scientific standards. Furthermore, in practice, many clinical ecologists de-emphasize possible psychosocial factors and psychological interventions, despite their stated concern about the *total* load of environmental stressors in illness. More conventional approaches, on the other hand, may overlook the chemical aspects of the environment and focus too heavily on psychogenic etiologies of illness. Clinical ecologists also note that psychosocial factors are often more difficult and time-consuming to test for and treat than are specific food and chemical sensitivities.

The work stimulated by clinical ecology may ultimately lead to a more balanced view of the role of all facets of the environment—including chemical and psychosocial[240]—in many disorders. If its basic tenets prove to be valid, clinical ecology would lead to significant advances in the understanding and treatment of chronic illnesses, as well as to drastic changes in eating and living styles. Consequently, this field merits a careful and objective evaluation.

RESOURCES

Books on Clinical Ecology for the Health Professional and Lay Reader

DICKEY, L.D. (ed.). *Clinical Ecology.* Springfield, Illinois: Charles C Thomas, 1976.

First compendium of writings by leading physicians and researchers in clinical ecology on the history, theories, clinical issues, and methods of diagnosis and treatment of environmentally-ill patients.

GERRARD, J.W. (ed.). *Food Allergies: New Perspectives.* Springfield, Illinois: Charles C Thomas, 1980.

Overview of the controversial field of food allergy, edited by a respected and experienced allergist.

MILLER, J.B. *Food Allergy: Provocative Testing and Injection Therapy.* Springfield, Illinois: Charles C Thomas, 1972.

Practical textbook by a leading allergist-pediatrician on the provocation-neutralization method. Gives a step-by-step approach for physicians and technicians.

O'BANION, D.R. *An Ecological and Nutritional Approach to Behavioral Medicine.* Springfield, Illinois: Charles C Thomas, 1981.

Integration of clinical ecology with issues in behavioral medicine by a clinical psychologist.

O'BANION, D.R. *The Ecological and Nutritional Treatment of Health Disorders.* Springfield, Illinois: Charles C Thomas, 1981.

Up-to-date presentation of clinical ecology theory and practice from the viewpoint of a clinical psychologist.

RANDOLPH, T.G. *Human Ecology and Susceptibility to the Chemical Environment.* Springfield, Illinois: Charles C Thomas, 1962 (Sixth printing 1978).
Original description by internist-allergist Randolph of his clinical observations and conclusions about chemically-sensitive patients.

RINKEL, H.J., RANDOLPH, T.G., AND ZELLER, M. *Food Allergy.* Norwalk, Connecticut: New England Foundation of Allergic and Environmental Disease (3 Brush St., Norwalk, Conn. 06850), reprint of 1951 text.
Classic textbook by pioneers in the field of food allergy, with particular attention to chronic adaptation phenomena and rotary diet approaches.

Lay Publications on Clinical Ecology

BELL, I.R. *EI: The Experience. Ecologic Illness from the Viewpoint of the Patient.* X-Press! Publishing (P.O. Box U, North Bend, WA 98045), 1982.

CROOK, W.G. *Can Your Child Read? Is He Hyperactive?* Jackson, Tenn: Pedicenter Press, 1975.

EAGLE R. *Eating and Allergy.* Garden City, New York: Doubleday and Co., 1981.

FORMAN, R. *How to Control Your Allergies.* New York: Larchmont Books, 1979.

GOLOS, N., GOLBITZ, F.G., LEIGHTON, F.S. *Coping with Your Allergies.* New York: Simon & Schuster, 1979.

GOLOS, N. AND GOLBITZ, F.G. *If This Is Tuesday, It Must Be Chicken or How to Rotate Your Food for Better Health.* Dallas: Human Ecology Research Foundation of Southwest (Webbs Chapel Rd., Dallas Tx. 75234), 1981.

LUDEMAN, K., HENDERSON, L., BASAYNE, H.S. *Do-It-Yourself Allergy Analysis Handbook.* New Canaan, Conn: Keat's Publishing, Inc., 1979.

MACKARNESS, R. *Chemical Victims.* London: Pan, 1980.

MACKARNESS, R. *Eating Dangerously: The Hazards of Hidden Allergies.* New York:Harcourt Brace Jovanovich, 1976.

MANDELL, M., SCANLON, L.W. *Dr. Mandell's 5-Day Allergy Relief System.* New York: Thomas Y. Crowell, 1979 (also in paperback, New York: Pocket Books, 1980).

MCGEE, C.T. *How to Survive Modern Technology.* Alamo, Calif: Ecology Press, 1979.

PHILPOTT, W.H. AND KALITA, D.K. *Brain Allergies: The Psycho-Nutrient Connection.* New Canaan, Conn: Keats Publishing, Inc., 1980.

RANDOLPH, T.G. AND MOSS, R.W. *An Alternative Approach to Allergies.* New York: Lippincott & Crowell, 1980.

RAPP, D.J. *Allergies and Your Family.* New York: Sterling Publishers, 1981.

RAPP, D.J. *Allergies and the Hyperactive Child.* New York: Cornerstone Library, 1981.

STATE OF CALIFORNIA. DEPARTMENT OF CONSUMER AFFAIRS. *Clean Your Room! A Compendium on Indoor Pollution.* (P.O. Box 310, 1020 North Street, Sacramento, CA 95802), 1982.

ZAMM, A.V. *Why Your House May Endanger Your Health .* New York: Simon and Schuster, 1980.

Popular Books on Nutrition with Particular Relevance to Clinical Ecology

HALL, R.H. *Food for Nought: The Decline in Nutrition.* New York: Vintage, 1974.

HUNTER, B.T. *The Great Nutrition Robbery.* New York: Charles Scribner's Sons, 1978.

NEWBOLD, H.L. *Mega-Nutrients for Your Nerves.* New York: Peter H. Wyden, 1975.

PFEIFFER, C.C. *Mental and Elemental Nutrients.* New Canaan, Conn: Keats Publishing, Inc., 1975.

WILLIAMS, R. *Biochemical Individuality.* New York: John Wiley & Sons, 1957.

WILLIAMS, R. *Nutrition Against Disease.* New York: Pitman, 1971.

Professional organization

Society for Clinical Ecology. Del Stigler, M.D., Secretary. 2005 Franklin Street, Suite 490, Denver, Colorado 80205.

Basic and advanced seminars on clinical and research areas; referral source for clinical ecology specialists; journal *Clinical Ecology* and newsletter publication.

Patient organization

Human Ecology Action League (HEAL). 505 North Lake Shore Drive, Suite 6506, Chicago, Illinois 60611.

National patient group with local chapters and national meetings; newsletter publication.

Dietary & Environmental Questionnaire

Name: Date:

Occupation:

Address:

Telephone: Age:

Height: Weight: Maximum Weight:

In answering the questions below, please be as specific and detailed as possible. For example, instead of simply listing "soda," give the particular brand (e.g., 7-Up or Coke). Please specify whether your cakes and breads are homemade or commercially-packaged. Also specify method of preparation or ingredients (e.g., Shake-and-Bake chicken, Colonel Sanders' or home-broiled without seasonings). Everything that you consume by mouth is important, including condiments such as spices, mustard, ketchup, salad dressing, etc. Even chewing gum and any artificial sweeteners that you use should be noted as a food. Please feel free to use an additional sheet or the backs of these pages for completing answers to any questions or for adding comments.

Part I. Diet

A. List everything eaten in the *last 24 hours.* Try to list by meals.

B. List your favorite foods and beverages, or those you particularly enjoy or crave.

C. Have you noticed that any particular foods *make you feel better* or give you a pick-up after eating them? Specify.

D. List all foods and beverages which you have ever *limited or avoided* by preference or necessity. If restriction was due to an adverse reaction, please specify nature of reaction.

E. Circle any of these foods which have ever made you feel ill in any way (not only those that cause gastrointestinal symptoms). Include those already given in question D.

almonds
cashews
peanuts
walnuts

clams
crab
shrimp
lobster
oysters
fish

strawberries
apples
bananas
dates
figs
coconut
oranges

tomatoes
potatoes
onion
cabbage
mushrooms
soybeans
peas

popcorn
buckwheat
mustard
cinnamon
nutmeg
garlic
carrots

chocolate
candy
cake
cookies
ice cream
bread

eggs
chicken
beef
pork
cheese
lamb

sausage
hot dog
salami
bologna
bacon

coffee
tea
milk
soda
alcoholic beverages

Chinese food
Italian food
Mexican food
French food
Indian food

other foods:

F. Do you take vitamin supplements? Specify type and amount per day (milligrams or units). Note whether "natural"—i.e., made from food products such as rose hips, citrus, yeast, wheat germ—or "synthetic."

G. Do you consume alcoholic beverages regularly? Specify types of beverage, amounts, and frequency of use, even if already listed above.

H. Have you ever noticed a marked change in occurrence of symptoms
(1) on a change in dietary habits (e.g., ulcer diet, hypoglycemia diet, binges of particular foods)? Specify.

(2) on changing your job, moving to a new home, or going on vacation? Specify.

I. Check the number of times each week that you eat or drink each of the following items. Please add any items to the list that are not already listed below.

Food Item	*Daily*	*3-6 times per week*	*1-2 times per week*	*Less often (specify frequency)*	*Never use*	*Comments*
Dairy Products:						
milk (fresh)						
milk (powdered)						
yogurt						
ice cream						
cottage cheese						
cream cheese						
American cheese						
cheddar cheese						
muenster cheese						
Swiss cheese						
other cheese (specify)						
butter						
eggs						
Grains						
bread, rolls, buns (white)						
bread, rolls, buns (whole wheat or rye)						
English muffins, bagels						
crackers						

Food Item	*Daily*	*3-6 times per week*	*1-2 times per week*	*Less often (specify frequency)*	*Never use*	*Comments*
spaghetti, macaroni, noodles, pasta						
pizza						
pancakes						
wheat cereal (hot or cold)						
oatmeal cereal						
corn cereal						
cookies, cakes						
malt						
malted milk						
corn sugar or corn syrup						
corn oil/margarine						
syrup in canned fruit						
corn, popcorn						
rice						
Legumes						
peanuts, peanut oil						
peas (green)						
lima beans						
green or string beans						
lentils						
soybeans, soy sauce, soy oil, tofu						
canned or baked beans						
other beans						
alfalfa (sprouts)						
Vegetables						
yams, sweet potatoes						
carrots						
celery						
tomatoes						
green peppers						
white potatoes						
potato chips						
red pepper (hot)						
cabbage						
broccoli						
cauliflower						
brussels sprouts						
lettuce						
spinach						
beets						
onion						
garlic						
asparagus						
squash						
pickles						
cucumbers						

Food Item	*Daily*	*3-6 times per week*	*1-2 times per week*	*Less often (specify frequency)*	*Never use*	*Comments*
mushrooms avocado						
Fruits pumpkin, including seeds melon lemon grapefruit oranges apples pears peaches apricots plums prunes cherries raspberries blackberries strawberries blueberries cranberries grapes raisins bananas pineapple dates figs						
Nuts & Seeds, Oils cashews almonds walnuts pecans pistachio nuts sunflower seeds, oil safflower oil cottonseed oil						
Meats, Poultry, Fish beef steaks, roasts, veal hamburgers pork chops, roasts bacon, sausage, ham hot dogs bologna, salami lamb chicken turkey codfish tuna salmon						

Food Item	*Daily*	*3-6 times per week*	*1-2 times per week*	*Less often (specify frequency)*	*Never use*	*Comments*
trout						
sole						
perch						
clams						
oysters						
shrimp, prawns						
Sweeteners & Sweets						
cane sugar						
beet sugar						
honey						
saccharin						
jelly or jam						
pancake syrup						
candy						
chocolate						
chewing gum						
Beverages						
coffee (regular)						
coffee (decaffeinated)						
artificial coffee creamer						
tea						
soda (regular)						
soda (diet)						
beer						
wine						
other alcoholic beverages						
Condiments & Spices						
vinegar						
mayonnaise						
ketchup						
mustard						
black or white pepper						
cinnamon						
nutmeg						

Part II. Environment

A. Do you smoke tobacco? If so, specify whether cigarettes, cigars or pipe, and how much per day. For how many years have you smoked?

B. Are you currently taking any medications (including oral contraceptives, aspirin or Tylenol, cold remedies, antihistamines), or regularly using any non-prescription drugs or recreational drugs (e.g., marijuana, cocaine, etc.)? Specify.

C. Have you ever experienced adverse reactions to any medications or drugs (e.g., aspirin, antihistamine, anesthetic, sulfa drugs, barbiturates)? Specify, including symptoms.

D. Are there any particular *odors or fumes* (indoors or outdoors) which you like, dislike, have felt better after smelling, or have felt worse after smelling? Specify the item(s) and the nature of your reaction or symptoms.

E. Rate your sense of smell:

- ☐ very acute
- ☐ normal
- ☐ poor
- ☐ absent

F. Rate your ability to detect leaking natural gas:

- ☐ very acute
- ☐ normal
- ☐ poor
- ☐ absent

G. Circle any of these items which have ever made you feel in any way physically or emotionally ill (include those already given in Part II, Question D). Put a mark (x) next to any items which make you feel better in any way or "high."

natural gas
gasoline or oil
automobile exhaust
nail polish or nail polish remover
dry-cleaned clothing or furnishings
fresh newspapers
household waxes
fumes from roof tar or road asphalt
household disinfectants, deodorants, detergents
rubbing alcohol
varnishes, paints
scented personal hygiene products (deodorants, aftershave, perfume, powders, hairspray)
turpentine
pine-scented products
dust
mold, mildew
pollens
cosmetics
moth balls
chlorinated water, pools
bleaches
fabric store odors
foam rubber pillows, mattresses, rug backings
tire store odors
plastic bags, food containers
interiors of new automobiles
insecticides
dyes
petroleum jelly
tobacco smoke
new carpet odors
new clothing odors
cats
dogs
other animals (specify)

H. List all products used for *personal hygiene,* such as specific brands of deodorants, after-shaves, perfumes, hair sprays, cosmetics, soaps, etc.

I. Check as applicable for your home environment:*

Home: How long have you lived there?____________________

☐ single house
☐ double house
☐ apartment
☐ hotel
☐ dormitory
☐ trailer

Region:
☐ city, residential area
☐ city, industrial area
☐ suburban
☐ small town
☐ rural

Garage:
☐ in separate unattached building
☐ with inside passageway to house
☐ in basement of house
☐ none

Heating and ventilation of home:
☐ gas
☐ oil
☐ coal
☐ warm air
☐ hot water or steam
☐ electric (heat pump)
☐ electric (radiant)
☐ space heaters
☐ fireplace or wood stove
☐ other

Air conditioning:
☐ window units
☐ central system

Filters:
☐ fiberglass
☐ oiled
☐ unoiled
☐ electrostatic
☐ activated carbon

Automobile air conditioning:
☐ factory-installed
☐ installed after car purchase

Kitchen exhaust:
Fan: ☐yes ☐no
Kitchen door:
☐ usually left open
☐ usually left closed

Range: age of unit____________
☐ electric
☐ gas
☐ oil

Refrigerator: age of unit____________
☐ electric
☐ gas

Clothes dryer: age of unit____________
☐ electric
☐ gas

Water heater: age of unit____________
☐ electric
☐ gas

Food storage:
☐ in glass
☐ in plastic
☐ in enamel ware
☐ in aluminum foil
☐ other ________________

Drinking water:
☐ spring or well
☐ softened
☐ chlorinated
☐ fluoridated

*These questions adapted from Randolph, T.G., *Human Ecology and Susceptibility to the Chemical Environment.* Springfield, Illinois: Charles C Thomas, 1962 (sixth printing 1978).

Furnishings and household maintenance:

Mattresses:
- ☐ cotton
- ☐ foam rubber
- ☐ plastic covered
- ☐ other ____________________

Pillows:
- ☐ feather
- ☐ foam rubber
- ☐ kapok
- ☐ Dacron
- ☐ plastic covered
- ☐ other ____________________

Carpeting or rugs:
- ☐ wall-to-wall
- ☐ scatter rugs
- ☐ linoleum
- ☐ bare wooden floors

Rug materials:
- ☐ wool
- ☐ cotton
- ☐ synthetic
- ☐ natural fiber
- ☐ rubber or plastic backing

Rug pads:
- ☐ plastic
- ☐ rubber
- ☐ hair

Upholstery coverings:
- ☐ cotton
- ☐ linen
- ☐ synthetic fabrics
- ☐ plastic
- ☐ wool
- ☐ silk

Upholstery padding:
check information on tags
- ☐ cotton
- ☐ hair
- ☐ rubber
- ☐ other ____________________

Curtains:
- ☐ cotton
- ☐ linen
- ☐ synthetic materials
- ☐ plastic
- ☐ wool
- ☐ silk

Furniture polish:
☐ yes ☐ no

Floor wax:
☐ yes ☐ no

Insect control: *specify type, brand*
- ☐ sprays
- ☐ no-pest strips
- ☐ moth crystals or moth balls
- ☐ exterminators

Pet animals:
breed: ____________________
- ☐ flea powder
- ☐ flea collar
- ☐ litter box

Laundry: *specify brands*
- ☐ soap
- ☐ bleaches
- ☐ ammonia
- ☐ detergents

J. Describe briefly the type of work you do, any regular chemical exposures of which you are aware, number of hours of work per day.

K. If you work outside your home, please answer the following questions about your work environment.

Work region:
- ☐ city
- ☐ suburban
- ☐ small town
- ☐ rural

Transportation to work:
- ☐ car
- ☐ bus
- ☐ train
- ☐ walking
- ☐ other ____________

Work setting:
- ☐ office
- ☐ factory
- ☐ store
- ☐ school
- ☐ farm
- ☐ truck or bus
- ☐ car or taxicab
- ☐ other (specify)____________

Is there a parking garage underneath or attached to your place of business? ☐ yes ☐ no

Distance between home and work:____________

Has your place of business been treated with pesticides?
☐ yes ☐ no ☐ don't know

Type of lighting:
- ☐ fluorescent (regular)
- ☐ fluorescent (full-spectrum)
- ☐ incandescent bulbs
- ☐ natural lighting (sunlight)
- ☐ windows present in work-room

Ventilation system:
- ☐ windows that can be opened
- ☐ windowless room
- ☐ fans
- ☐ central ventilation system
- ☐ air conditioning (window units)
- ☐ air conditioning (central system)

Is there usually tobacco smoke in or near your work area?
☐ yes ☐ no

Heating system:

Fuel:
- ☐ gas
- ☐ oil
- ☐ electric
- ☐ coal
- ☐ other (specify)____________

Type:
- ☐ warm air
- ☐ hot water or steam
- ☐ electric
- ☐ space heaters
- ☐ fireplace or wood stove

Floor covering:
- ☐ wall-to-wall carpets
- ☐ scatter rugs
- ☐ linoleum
- ☐ bare wooden floors
- ☐ other (specify)____________

Rug materials:
- ☐ wool
- ☐ cotton
- ☐ synthetic
- ☐ natural fiber
- ☐ rubber or plastic backing

Curtains: specify material if known___________________________

Upholstery: specify material if known___________________________

List the types of machines or equipment which are present in or near your immediate work area. (Include typewriters, word-processors, computers, copying machines, refrigeration units, stoves, dishwashers, X-ray equipment, audio or visual electronics, cash registers, etc.)

R E F E R E N C E S

1. RANDOLPH, T.G. *Human Ecology and Susceptibility to the Chemical Environment.* Springfield, Illinois: Charles C Thomas, 1962 (sixth printing 1978).
2. DICKEY, L.D., (ED.) *Clinical Ecology.* Springfield, Illinois: Charles C Thomas, 1976.
3. ADELMAN, R.C. Loss of adaptive mechanisms during aging. *Fed Proc.* 38:1968-1971, 1979.
4. CALABRESE, E.J. *Pollutants and High-Risk Groups. The Biological Basis of Increased Human Susceptibility to Environmental and Occupational Pollutants.* NY: Wiley-Interscience, 1977.
5. CALABRESE, E.J. Animal model of human disease. Increased sensitivity to ozone. Mice with low levels of G-6-PD. *Amer J Pathol.* 91:409-411, 1978.
6. BAFITIS, H., SMOLENSKY, M.H., BARTHOLOMEW, P.H., ET AL. A circadian susceptibility/resistance rhythm for potassium cyanide in male BALB/cCr mice. *Toxicology.* 11:251-258, 1978.
7. SHAKMAN, R.A. Nutritional influences on the toxicity of environmental pollutants. *Arch Environ Health.* 28:105-113, 1974.
8. WEINSIER, R.L., HUNKER, E.M., KRUMDIECK, C.L., ET AL. Hospital malnutrition: a prospective evaluation of general medical patients during the course of hospitalization. *Am J Clin Nutr.* 32:418-426, 1979.
9. RANDOLPH, T.G. Specific adaptation. *Ann Allergy.* 40:333-45, 1978.
10. SAVOLAINEN, H., PFAFFLI, P., HELOJOKI, M., AND TENGEN, M. Neurochemical and behavioral effects of long-term intermittent inhalation of xylene vapour and simultaneous ethanol intake. *Acta Pharmacol et Toxicol.* 44:200-07, 1979.
11. HANNINEN, H., ESKELINEN, L., HUSMAN, K., AND NURMINEN, M. Behavioral effects of long-term exposure to a mixture of organic solvents. *Scand J Work Environ Health.* 4:240-55, 1976.
12. ELOFSSON, S.A., GAMBERALE, F., HINDMARSH, T., ET AL. Exposure to organic solvents. A cross-sectional investigation on occupationally exposed car and industrial spray painters with special reference to the nervous system. *Scand J Work Environ Health.* 6:239-73, 1980.
13. WHITE, J. F. AND CARLSON, G.P. Epinephrine-induced cardiac arrhythmias in rabbits exposed to trichloroethylene: potentiation by ethanol metabolites. *Toxicol Appl Pharmacol.* 60:466-71, 1981.
14. CONNELL, JOHN. Quantitative intranasal pollen challenges III. The priming effect in allergic rhinitis. *J Allergy.* 43:33-44, 1969.
15. BRUHN, P., ARLIEN-SBORG, P., GYLDENSTED, C., ET AL. Prognosis in chronic toxic encephalopathy. A two-year follow-up study in 26 house painters with occupational encephalopathy. *Acta Neurol Scand.* 64:259-72, 1981.
16. DOHAN, F.C., GRASBERGER, J.C., LOWELL, F.M., ET AL. Relapsed

schizophrenics: more rapid improvement on a milk and cereal-free diet. *Br J Psychiat.* 115:595-596, 1969.
17. DOHAN, F.C. AND GRASBERGER, J.C. Relapsed schizophrenics: earlier discharge from the hospital after cereal-free, milk-free diet. *Amer J Psychiat.* 130:685-688, 1973.
18. SINGH, M.M. AND KAY, S.R. Wheat gluten as a pathogenic factor in schizophrenia. *Science.* 191:401-402, 1976.
19. POTKIN, S.G., WEINBERGER, D., KLEINMAN, J., ET AL Wheat gluten challenge in schizophrenic patients. *Amer J Psychiat.* 138:1208-1211, 1981.
20. FEINGOLD, B.F. *Why Your Child Is Hyperactive.* NY: Random House, 1975.
21. CONNERS, C.K., GOYETTE, C.H., AND NEWMAN, E.B. Dose-time effect of artificial colors in hyperactive children. *J Learning Disabilities.* 13:48-52, 1980.
22. HARLEY, J.P., MATTHEWS, C.G., AND EICHMAN, P. Synthetic food colors and hyperactivity in children: a double-blind challenge experiment. *Pediatrics.* 62:975-983, 1978.
23. SWANSON, J.M. AND KINSBOURNE, M. Food dyes impair performance of hyperactive children on a laboratory learning test. *Science.* 207:1485-1486, 1980.
24. WEISS, B., WILLIAMS, J.H., MARGEN, S., ET AL. Behavioral responses to artificial food colors. *Science.* 207:1487-1488, 1980.
25. PRINZ, R.J., ROBERTS, W.A., AND HANTMAN, E. Dietary correlates of hyperactive behavior in children. *J Consulting Clin Psychol.* 48:760-769, 1980.
26. GARDNER, D.E., LEWIS, T.R., ALPERT, S.M., ET AL. The role of tolerance in pulmonary defense mechanisms. *Arch Environ Health.* 25:432-438, 1972.
27. GOLDMAN, R.H. AND PETERS, J.M. The occupational and environmental health history. *JAMA.* 246:2831-36, 1981.
28. HACKNEY, J.D., LINN, W.S., BUCKLEY, R.D., ET AL. Respiratory and biochemical adaptations in men repeatedly exposed to ozone. In Folinsbee, L.J., Wagner, J.A., Borgia, J.F. et al.(eds). *Environmental Stress. Individual Human Adaptations.* NY: Academic Press, 1978, pp 111-124.
29. HACKNEY, J.D., LINN, W.S., BUCKLEY, R.D., ET AL. Studies in adaption to ambient oxidant air pollution: effects of ozone exposure in Los Angeles residents vs. new arrivals. *Environ Health Perspect.* 18:141-146, 1976.
30. HACKNEY, J.D., LINN, W.S., KARUZA, S.K., ET AL. Effects of ozone exposure in Canadians and Southern Californians. *Arch Environ Health.* 32:110-16, 1977.
31. HACKNEY, J.D., LINN, W.S., MOHLER, J.G., ET AL. Adaptation to short-term respiratory effects of ozone in men exposed repeatedly. *J Appl Physiol.* 43:82-85, 1977
32. FAIRCHILD, E.J. Tolerance mechanisms. Determinants of lung responses to injurious agents. *Arch Environ Health.* 14:111-126, 1967.
33. National Research Council. Committee on Medical and Biologic Effects of Environmental Pollutants. *Ozone and Other Photochemical Oxidants.* Washington: Nat Acad Sci, 1977 (ISBN-0-309-02531-1)
34. HACKNEY, J.D., LINN, W.S., LAW, D.C., ET AL. Experimental studies on human health effects of air pollutants. III. Two-hour exposure to ozone alone and in combination with other pollutant gases. *Arch Environ Health.* 30:385-390, 1975.
35. STOKINGER, H.E., WAGNER, W.D., AND DOBROGORSKI, O.J. Ozone toxicity studies. III. Chronic injury to lungs of animals following exposure at a low level. *Arch Indust Health.* 16:514-522, 1957.
36. FOLINSBEE, L.J., BEDI, J.F., AND HORVATH, S.M. Respiratory responses in humans repeatedly exposed to low concentrations of ozone. *Amer Rev Resp Dis.* 121:431-39, 1980.
37. LOVE, G.J., LAN, S.P., SHY, C.M., ET AL.The incidence and severity of acute respiratory illness in families exposed to different levels of air pollution, N.Y. metropolitan area, 1971-1972. *Arch Environ Health.* 36:66-74, 1981.
38. DIMEO, M.J., GLENN, M.G., HOLTZMAN, M.J. ET AL. Threshold concentration of ozone causing an increase in bronchial reactivity in humans and adaptation with repeated exposures. *Am Rev Respir Dis.* 124:245-248, 1981.
39. RINKEL, H.J., RANDOLPH, T.G., AND ZELLER, M. *Food Allergy.* Springfield, Illinois, 1951. (Reprinted by N.E. Foundation for Allergic and Environmental Diseases, Norwalk, Conn.)
40. GARDNER, D.E., COFFIN, D.L., PINIGIN, M.A. ET AL. Role of time as a factor in the toxicity of chemical compounds in intermittent and continuous exposures. Part I. Effects of continuous exposure. *J Toxicol Environ Health* 3:811-820, 1977.
41. COFFIN, D.L., GARDNER, D.E., SIDORENKO, G.I., ET AL. Role of time as a factor in the toxicity of chemical compounds in intermittent and continuous exposures. Part II. Effects of intermittent exposure. *J Toxicol Environ Health. 3:821-828, 1977.*
42. RAPP, D.J. Does diet affect hyperactivity? *J Learning Disabilities.* 11:56-62, 1978.

43. Rea, W.J. Environmentally triggered cardiac disease. *Ann Allergy.* 40:243-51, 1978.
44. Rea, W.J. Environmentally triggered small vessel vasculitis.*Ann Allergy.* 38:245-51,1977.
45. Rea, W.J. Environmentally triggered thrombophlebitis. *Ann Allergy.* 37:101-109, 1976.
46. Rea. W.J., Bell, I.R., Suits, C.W., et al. Food and chemical susceptibility after environmental chemical overexposure: case histories. *Ann Allergy.* 41:101-10, 1978.
47. Sandberg, D.H., McIntosh, R.M., Bernstein, C.W., et al. Severe steroid-responsive nephrosis associated with hypersensitivity. *Lancet.*1:388-90, 1977.
48. Rapp, D.J. Double-blind confirmation and treatment of milk sensitivity. *Med J Aust.* 1:571-572,1978.
49. O'Banion, D., Armstrong, B., Cummings, R.A., et al. Disruptive behavior: a dietary aproach. *J Autism Childhood Schizophrenia.* 8:325-337, 1978.
50. Hare, F. *The Food Factor in Disease.* London: Longmans and Green, 1905.
51. Rowe, A.H. *Food Allergy. Its Manifestations, Diagnosis, and Treatment.* Philadelphia: Lea and Febiger, 1931.
52. Rowe, A.H. and Rowe, A. Jr. *Food Allergy. Its Manifestations and Control and the Elimination Diets. A Compendium.* Springfield, Illinois: Charles C Thomas, 1972.
53. Rinkel, H.J. Food allergy. II. The technique and clinical application of individual food tests. *Ann Allergy.* 2:504-514, 1944.
54. Randolph, T.G. The descriptive features of food addiction. Addictive eating and drinking. *Q J Stud Alc.* 17:198-224, 1956.
55. Goldstein, A., Aronow, L., Kalman, S.M. *Principles of Drug Action: The Basis of Pharmacology.* NY: John Wiley and Sons, 1974, pp 569-621.
56. Klee, W.A., Zioudrou, C., and Streaty, R.A. Exorphins: peptides with opioid activity isolated from wheat gluten, and their possible role in the etiology of schizophrenia. In Usdin, E., Bunney, W.E. Jr., and Kline, N.S. (eds): *Endorphins in Mental Health Research.* NY: Oxford Univ. Press, 1979, pp 209-18.
57. Zetler, G. Analgesia and ptosis caused by caerulein and CCK octapeptide (CCK-8). *Neuropharmacol.* 19:415-422, 1980.
58. Oehme, P., Hilse, H., Morgenstern, E., et al. Substance P: does it produce analgesia or hyperalgesia? *Science.* 208:305-307, 1980 (Apr 18).
59. Akande, B., Reilly, P., Modlin, I.M., et al. Radioimmunoassay measurement of substance P release following a meat meal. *Surgery.* 89:378-83, 1981.
60. Brantl, V. and Teschemacher, H. A material with opioid activity in bovine milk and milk products. *Naunyn Schmiedebergs Arch Pharmacol.* 306:301-304, 1979.
61. Kariya, K., Okamato, H., Iwaki, H., et al. Studies on the effects of bradykinin and its fragments on the central nervous system. *Adv Exp Med Biol.* 120A:487-97, 1979.
62. DaSilva, G.R. and Rocha e Silva, M. Catatonia induced in the rabbit by intracerebral injection of bradykinin and morphine. *Eur J Pharmacol.* 15:180-186, 1971.
63. Bell, I.R. A kinin model of mediation for food and chemical sensitivities: biobehavioral implications. *Ann Allergy.* 35:206-15, 1975.
64. Rinkel, H.J. Food allergy IV. The function and clinical application of the rotary diversified diet. *J Pediat.* 32:266, 1948.
65. Wilson, R. Risks caused by low levels of pollution. *Yale J Biol Med.* 51:37-51, 1978.
66. Randolph, T.G. Ecologic orientation in medicine: comprehensive environmental control in diagnosis and therapy. *Ann Allergy.* 23:7-22, 1965.
67. Edgar, R.T., Fenyves, E.J., and Rea, W.J. Air pollution analysis used in operating an environmental control unit. *Ann Allergy.* 42:166-173, 1979.
68. Kon, S.H. Underestimation of chronic toxicities of food additives and chemicals: the bias of a phantom rule. *Med Hypotheses.* 4:324-339, 1978.
69. Randolph, T.G. The history of ecologic mental illness. In Frazier, C.A. (ed) *Annual Rev Allergy 1973.* Flushing, NY: Med Exam Pub Co, 1974, pp 425-441.
70. Knave, B., Mindus, P., and Struwe, G. Neurasthenic symptoms in workers occupationally exposed to jet fuel. *Acta Psychiat Scand.* 60:39-49, 1979.
71. King, D.S. Can allergic exposure provoke psychological symptoms? A double-blind test. *Biol Psychiat.* 16:3-19, 1981.
72. Savage, G.H. *Insanity and Allied Neuroses.* Philadelphia: Lea and Febiger, 1884.
73. Baldwin, J.A. Schizophrenia and physical disease. *Psychol Medicine.* 9:611-618, 1979.
74. Davis, D. and Offenkranz, W. Is there a reciprocal relationship between symptoms and affect in asthma? *J Nerv Ment Dis.* 163:369-389, 1976.
75. Rowe, A.H. Allergic toxemia and migraine due to food allergy. *Calif West Med.* 33:785-792, 1930.
76. Randolph, T.G. Fatigue and weakness of

allergic origin (allergic toxemia) to be differentiated from "nervous fatigue" or neurasthenia. *Ann Allergy. 3:418-430, 1945.*
77. SPEER, F. The allergic tension-fatigue syndrome. *Ped Clin N Amer.* 1:1029-1037, 1954.
78. CHERRY, N., WALDRON, H.A., WELLS, G.G., ET AL. An investigation of the acute behavioral effects of styrene on factory workers. *Br J Ind Med.* 37:234-240, 1980.
79. GOODWIN, D.W. AND GUZE, S.B. *Psychiatric Diagnosis.* (2nd ed) NY: Oxford University Press, 1979, pp. 51-69.
80. KNAVE, B., PERSSON, H.E., GOLDBERG, J.M., ET AL. Long-term exposure to jet fuel. An investigation on occupationally exposed workers with special reference to the nervous system. *Scand J Work Environ Health.* 32:152-164, 1976.
81. FAUST, H.S. AND BRILLIANT, L.B. Is the diagnosis of "mass hysteria" an excuse for incomplete investigation of low-level environmental contamination? *J Occupat Med.* 23:22-26, 1981.
82. PERLEY, M.J. AND GUZE, S.B. Hysteria—the stability and usefulness of clinical criteria. *NEJM.* 266:421-426, 1962.
83. HANE, M., AXELSON, D., BLUME, J., ET AL. Psychological function changes among house painters. *Scand J Work Environ Health.* 3:91-99, 1977.
84. RAVNSKOV, U., FORSBERG, B., AND SKERFVING, S. Glomerulonephritis and exposure to organic solvents. *Acta Med Scand.* 205:575-579, 1979.
85. FINN, R., FENNERTY, A.G., AND AHMAD, R. Hydrocarbon exposure and glomerulonephritis. *Clin Nephrol.* 14:173-175, 1980.
86. VAN DER LAAN, G. Chronic glomerulonephritis and organic solvents. A case-control study. *Int Arch Occup Environ Health.* 47:1-8, 1980.
87. BAHN, A.K., MILLS, J.L., SNYDER, P.J., ET AL. Hypothyroidism in workers exposed to polybrominated biphenyls. *NEJM.* 302:31-33, 1980.
88. ZELLER, M. Rheumatoid arthritis—food allergy as a factor. *Ann Allergy.* 7:200-205, 1949.
89. BALYEAT, R.M. AND BRITTAIN, F.L. Allergic migraine. Based on the study of 55 cases. *Amer J Med Sci.* 180:212-221, 1930.
90. MONRO, J., CARINI, C., BROSTOFF, J., ET AL. Food allergy in migraine. Study of dietary exclusion and RAST. *Lancet.* 2:1-4, 1980.
91. ROWE, A.H. AND ROWE, A.,JR. Chronic ulcerative colitis and regional enteritis—their allergic aspects. *Ann Allergy.* 12:387-402, 1954.
92. ALEXANDER, F. *Psychosomatic Medicine.* NY: W.W. Norton, 1950.
93. RAPP, D.J. Double-blind confirmation and treatment of milk sensitivity. *Med J Aust.* 1:571-572, 1978.
94. REA, V.J., ET AL. Food and chemical susceptibility after environmental chemical overexposure—case histories. *Ann Allergy.* 41:101-110, 1978.
95. STOYVA, J. Why should muscular relaxation be clinically useful? Some data and 2½ models. In Beatty, J. and Legewie, H. (eds). *Biofeedback and Behavior.* NY: Plenum, 1977, pp 449-72.
96. ADER, R. (ED). *Psychoneuroimmunology.* NY: Academic Press, 1981.
97. SOLOMON, G.F. AND AMKRAUT, A.A. Psychoneuroendocrinological effects on the immune response. *Ann Rev Microbiol.* 35:155-184, 1981.
98. DUKOR, P., KALLOS, P., SCHLUMBERGER, H.D., ET AL. (EDS.) PAR. *Pseudo-Allergic Reactions. Involvement of Drugs and Chemicals.* Vol. 1. Genetic Aspects and Anaphylactoid Reactions; Vol. 2. Cytotoxic and Complement Mediated Reactions; Vol. 3. Cell Mediated Reactions. Basel: S. Karger, 1981.
99. NEUMAN, I., ELIAN, H., NAHUM, H., ET AL. The danger of 'yellow dyes' (tartrazine) to allergic subjects. *Clin Allergy.* 8:65-68, 1978.
100. NEUMAN, I. AND CRETER, D. Hereditary complement defect and kinins involvement in asthma. *Ann Allergy.* 38:79-82, 1977.
101. MCGOVERN, J.J. Correlation of clinical food allergy symptoms with serial pharmacological and immunological changes in the patient's plasma. *Ann Allergy.* 44:57-58, 1980 (abstract).
102. MCGOVERN, J.J. Blood chemistry abnormalities in patients with food and chemical sensitivities. *Ortho Med.* 5:13, 1980 (abstract).
103. LEVIN, A.S., MCGOVERN, J.J., MILLER, J.B., ET AL. Immune complex mediated vascular inflammation in patients with food and chemical allergies. *Ann Allergy.* 47:138, 1981 (abstract).
104. ASHKENAZI, A., LEVIN, S., IDAR, D., ET AL. In vitro cell-mediated immunologic assay for cow's milk allergy. *Pediatrics.* 66:399-402, 1980.
105. TREVINO, R. J. Immunologic mechanisms in the production of food sensitivities. *Laryngoscope.* 91:1913-36, 1981.
106. GALLAGHER, J.S., TSE, C.S., BROOKS, S.M., ET AL. Diverse profiles of immunoreactivity in toluene diisocyanate

(TDI) asthma. *JOM.* 23:610-16, 1981.

107. Paganelli, R., Levinsky, R. J., Atherton, D.J. Detection of specific antigen within circulating immune complexes: validation of the assay and its application to food antigen-antibody complexes formed in healthy and food-allergic subjects. *Clin Exp Immunol.* 46:44-53, 1981.
108. Valverde, E., Vich, J.M., Garcia-Calderon, J.V., et al. In vitro stimulation of lymphocytes in patients with chronic urticaria induced by additives and food. *Clin Allergy.* 10:691-98, 1980.
109. Galant, S.P., Bullock, J., and Frick, O.L. An immunological approach to the diagnosis of food sensitivity. *Clin Allergy. 3:363-372, 1973.*
110. Minor, J.D., Tolber, S.G., and Frick, O.L. Leukocyte inhibition factor in delayed-onset food allergy. *J Allergy Clin Immunol.* 66:314-321, 1980.
111. Coombs, R.R.A., and Oldham, G. Early rheumatoid-like joint lesions in rabbits drinking cows' milk. *Int Archs Allergy Appl Immunol.* 64:287-292, 1981.
112. Oldham, G. and Coombs, R.R.A. Early rheumatoid-like lesions in rabbits injected with foreign serum or milk proteins. III. Influence of concommitant IgE-like antibodies and of the breed of rabbit. *Int Archs Allergy Appl Immunol. 61:81-90, 1980.*
113. Poole, A. R., Oldham, G., and Coombs, R.R.A. Early rheumatoid-like lesions in rabbits injected with foreign serum: relationship to localization of immune complexes in the lining tissues of joints and cellular content of synovial fluid. *Int Archs Allergy Appl Immunol.* 57:135-145, 1978.
114. Poole, A.R. and Coombs, R.R.A. Rheumatoid-like joint lesions in rabbits injected intravenously with bovine serum. *Int Archs Allergy Appl Immunol.* 54:97-113, 1977.
115. Malinow, M.R., Bardana, E.J., Pirofsky, B., et al. Systemic lupus erythematosus-like syndrome in monkeys fed alfalfa sprouts: role of a nonprotein amino acid. *Science.* 216:415-17, 1982.
116. Ward, A.M. Evidence of an immune complex disorder in vinyl chloride workers. *Proc Roy Soc Med.* 69:289-290, 1976 (abstract).
117 Harbeck, R.J., Hoffman, A.A., Hoffman, S.A., et al. Cerebrospinal fluid and the choroid plexus during acute immune complex disease. *Clin Immunol Immunopathol.* 13:413-425, 1979.
118. Hoffman, S.A., Shucard, D.W., Harbeck, R.J., et al. Chronic immune complex disease: behavioral and immunological correlates. *J Neuropathol Exp Neurol.* 37:426-436, 1978.
119. Bell, I.R., Guilleminault, C., and Dement, W.C. Hypersomnia, multiple-system symptomatology, and selective IgA deficiency. *Biol Psychiat.* 13:751-57, 1978.
120. Cunningham-Rundles, C., Brandeis, W.E., Pudifin, D.J. et al. Autoimmunity in selective IgA deficiency: relationship to anti-bovine protein antibodies, circulating immune complexes and clinical disease. *Clin Exp Immunol.* 45:299-304, 1981.
121. Teshima, H., Inoue, S., Ago, Y., et al. Plasminic activity and emotional stress. *Psychother Psychosom.* 23:218-228, 1974.
122. Thonnard-Nenmann, E. and Neckers, L.M. T-lymphocytes in migraine. *Ann Allergy.* 47:325-27, 1981.
123. Rivlin, J., Kuperman, O., Freier, S., et al. Suppressor T-lymphocyte activity in wheezy children with and without treatment by hyposensitization. *Clin Allergy.* 11:353-56, 1981.
124. Martinez, J.D., Santos, J., Stechschutte, D.J., et al. Nonspecific suppressor cell function in atopic subjects. *J Allergy Clin Immunol.* 64:485-90, 1979.
125. Valverde, E., Vich, J.M., Garcia-Calderon, J.V., et al. In vitro response of lymphocytes in patients with allergic tension-fatigue syndrome. *Ann Allergy.* 45:185-88, 1980.
126. Peterson, M.L., Rommo, N., House, D., et al. In vitro responsiveness of lymphocytes to phytohemmagglutinin. *Arch Environ Health.* 33:59-63, 1978.
127. Asher, I.M. (ed) *Inadvertent Modification of the Immune Response. The Effects of Foods, Drugs, and Environmental Contaminants.* Proceedings of the Fourth FDA Science Symposium, 1978 (U.S. Government Printing Office #017-012-00286-2).
128. Dean, J.H. and Padarathsingh, M.L. (eds) *The Biological Relevance of Immune Suppression Induced by Therapeutic and Environmental Chemicals.* NY: Van Vostrand Reinhold Co., 1981.
129. Archer, D.L., Bukovic-Wess, J.A., and Smith, B.G. Suppression of macrophage-dependent T-lymphocyte function(s) by gallic acid, a food additive metabolite. *Proc Soc Exp Biol Med.* 156:465-69, 1977.
130. Blalock, J.E., Archer, D.L., and Johnson, H.M. Anticellular and im-

munosuppressive activities of foodborne phenolic compounds. *Proc Soc Exp Biol Med.* 167:391-93, 1981.

131. Archer, D.L., Bukovic-Wess, J.A., and Smith, B.G. Inhibitory effect of an antioxidant, butylated hydroxyanisole, on the primary in vitro immune response. *Proc Soc Exp Biol Med.* 154:289-94, 1977.
132. Catanzaro, P.J., Schwartz, H.J., and Graham, R.C.,Jr. Spectrum and possible mechanism of carrageenan cytotoxicity. *Amer J Pathol.* 64:387-99, 1971.
133. Street, J.C. and Sharma, R.P. Alteration of induced cellular and humoral immune responses by pesticides and chemicals of environmental concern: quantitative studies of immunosuppression by DDT, Arochlor 1254, Carbaryl carbofuran, and methylparathion. *Toxicol Appl Pharmacol.* 32:587-602 , 1975.
134. Faith, R.E., Luster, M.I., and Vos, J.G. Effects on immunocompetence by chemicals of environmental concern. *Rev Biochem Toxicol.* 2:173-211, 1980.
135. Keller, S.E., Weiss, J.M., Schleifer, S.J., et al . Suppression of immunity by stress: effect of a graded series of stressors on lymphocyte stimulation in the rat. *Science.* 213:1397-99, 1981.
136. Bartrop, R.W., Lazarus, L., Luckhurst, E., et al. Depressed lymphocyte function after bereavement. *Lancet.* 1:834-836, 1977.
137. Buisseret, P.D., Youlten, L.J.F., Heinzelmann, D.I., et al. Prostaglandin-synthesis inhibitors in prophylaxis of food intolerance. *Lancet.* 1:906-908, 1978.
138. Pasculescu, G.L., Toma, C., Ghircoiasu, T., et al. Kallicreinemia and serotoninemia in patients with lactose malabsorption. *Med Interna (Bucur).* 26:601-610, 1974.
139. Berteau, P.E. and Deen, W.A. Changes in whole blood serotonin concentrations in rats exposed to insecticide aerosols. *Toxicol Appl Pharmacol.* 37:134, 1976 (abstract).
140. Sicuteri, F. Phenomenal similarities of migraine and morphine abstinence. *Headache.* 19:232-33, 1979.
141. Peltola, P. and Leppanen, V. The possible role of serotonin in neurocirculatory asthenia and functional cardiac disorders. *Annales Med Intern Fenn.* 52:21-33, 1963.
142. Bellanti, J.A., Nerarkar, L.S., and Willoughby, J.W. Measurement of plasma histamine in patients with suspected food hypersensitivity. *Ann Allergy.* 47:260-63, 1981.
143. Johnson, A.R. and Erdos, E.G. Release of histamine from mast cells by vasoactive peptides. *Proc Soc Exp Biol Med.* 142:1252-56, 1973.
144. Levin H.S. and Rodnitzky, R.L. Behavioral effects of organophosphate pesticides in man. *Clin Toxicol.* 9:391-405, 1976.
145. Thomas, G. and West, G.B. Prostaglandins as regulators of bradykinin responses. *J Pharm Pharmacol.* 25:747-748, 1973.
146. Wybran, J., Appelboom, T., Famaey, J.P., et al. Suggestive evidence for receptors for morphine and methionine-enkephalin on normal human blood T lymphocytes. *J Immunol,* 123:1068-70, 1979.
147. Plotnikoff, N.P. The central nervous system control of the immune system. Enkephalins: antitumor activities. Annual Meeting of the American College of Neuropsychopharmacology, San Diego, Dec, 1981 (abstract, p.67).
148. Longacre, S.L., Kocsis, J.J., Snyder, R. Influence of strain differences in mice on the metabolism and toxicity of benzene. *Toxicol Appl Pharmacol.* 60:398-409, 1981.
149. Gross, R.L. and Newberne, P. M. Role of nutrition in immunologic function. *Physiol Rev.* 60:188-302, 1980.
150. Keller, S.E., Stein, M., Camerino, M.S., et al. Suppression of lymphocyte stimulation by anterior hypothalamic lesions in the guinea pig. *Cell Immunol.* 52:334-40, 1980.
151. Schiavi, R.C., Macris, N.T., Camerino, M.S., et al.Effect of hypothalamic lesions on immediate hypersensitivity. *Amer J Physiol.* 228:596-601, 1975.
152. Schiavi, R.C., Adams. J., and Stein, M. Effect of hypothalamic lesions on histamine toxicity in the guinea pig. *Amer J Physiol.* 211:1269-1273, 1966.
153. Vasilieva, G.K, Frolov, E.P., Shatilova, N.V., et al. The effect of stimulation of the hypothalamus on kininogen dynamics in the blood serum of rabbits during sensitization and anaphylactic shock. *Biull Ekxp Biol Med.* 76:28, 1973 (Eng. abstract).
154. Besedovsky, H., Sorkin, E., Felix, D., et al. Hypothalamic changes during the immune response. *Eur J Immunol.* 7:323-325, 1977.
155. Kerr, F.W. and Pozuelo, J. Suppression of physical dependence and induction of hypersensitivity to morphine by stereotoxic hypothalamic lesions in addicted rats. A new theory of addiction. *Mayo Clin Proc.* 46:653-665, 1971.
156. Kellar, K.J., Langley, A.E., Marks, B.H., et al. Ventral medial

hypothalamus: involvement in hypoglycemic convulsions. *Science.* 187:746-748, 1975.
157. HALL, K. Allergy of the nervous system: a review. *Ann Allergy.* 36:49-64, 1976.
158. WURTMAN, R.J. AND FERNSTROM, J.D. Control of brain monoamine synthesis by diet and plasma amino acids. *Amer J Clin Nut.* 28:638-647, 1975.
159. RAPAPORT, S.I., KLEE, W.A., PETTIGREW, K D., ET AL. Entry of opioid peptides into the central nervous system. *Science.* 207:84-86, 1979.
160. BAERTSCHI, A.J., ZINGG, H.H., AND DREIFUSS, J.J. Enkephalins, substance P, bradykinin, and angiotensin II: Differential sites of action on the hypothalamo-neurohypophysial system. *Brain Res.* 220:107-119, 1981.
161. DAMSTRA, T. Environmental chemicals and nervous system dysfunction. *Yale J Biol Med.* 51:457-468, 1978.
162. HEMMINGS, W.A. Entry into brain of large molecules derived from dietary protein. Proceedings of Royal Society of London. 200:175-192, 1978.
163. MYERS, R.D. AND MCCALEB, M.L. Feeding: satiety signal from intestine triggers brain's noradrenergic mechanism. *Science.* 209:1035-1037, 1980.
164. POWLEY, T.L. AND LAUGHTON, W. Neural pathways involved in the hypothalamic integration of autonomic responses. *Diabetologia.* 20 (suppl):378-387, 1981.
165. KOMISARUK, B.R. AND BEYER, C. Responses of diencephalic neurons to olfactory bulb stimulation, odor, and arousal. *Brain Res.* 36:153-170, 1972.
166. LEONARD, B.E. AND TUITE M. Anatomical, physiological, and behavioral aspects of olfactory bulbectomy in the rat. *Int Rev Neurobiol.* 22:251-286, 1981.
167. JACKSON, R.T., TIGGES, J., AND ARNOLD, W. Subarachnoid space of the CNS, nasal mucosa, and lymphatic system. *Arch Otolaryngol.* 105:180-4, 1979.
168. BOKINA, A.I., EKSLER, N.D., SEMENENKO, A.D., ET AL. Investigation of the mechanism of action of atmospheric pollutants on the central nervous system and comparative evaluation of methods of study. *Environ Health Perspectives.* 13:37-42, 1976.
169. BELL, I.R. Food and chemical factors in sleep disorders. Presented at Tenth Advanced Seminar in Clinical Ecology. Dallas, Texas, December, 1976.
170. WOOD, R.W. Stimulus properties of inhaled substances. *Environ Health Perspect.* 26:69-76, 1978.
171. SIEGEL, S. Morphine analgesic tolerance: its situation specificity supports a Pavlovian conditioning model. *Science.* 193:323-25, 1976.
172. SIEGEL, S., HINSON, R.E., KRONK, M.D., ET AL. Heroin "overdose" death: contribution of drug-associated environmental cues. *Science.* 216:436-37, 1982.
173. ADER, R. AND COHEN, N. Behaviorally conditioned immunosuppression. *Psychosom Med.* 37:333-40, 1975.
174. CROOK, W.G., HARRISON, W.W., CRAWFORD, S.E., ET AL. Systemic manifestations due to allergy. *Pediatrics.* 27:790-799, 1961.
175. MELIA, R.J.W., FLOREY, C., ALTMAN, D.G., ET AL. Association between gas cooking and respiratory disease in children. *Br Med J.* 2:149-152, 1977.
176. MELIA, R.J.W., FLOREY, C., AND CHINN, S. The relation between respiratory illness in primary schoolchildren and the use of gas for cooking. I. Results from a national survey. *Internat J Epidemiol.* 8:333-338, 1979.
177. GOLDSTEIN, B.D., MELIA, R.J.W., CHINN, S., ET AL. The relation between respiratory illness in primary schoolchildren and the use of gas for cooking. II. Factors affecting nitrogen dioxide levels in the home. *Int J Epidemiol.* 8:339-345, 1979.
178. FLOREY, C., MELIA, R.J.W., CHINN, S., ET AL. The relation between respiratory illlness in primary schoolchildren and the use of gas for cooking. III. Nitrogen dioxide, respiratory illness, and lung infection. *Int J Epidemiol.* 8:347-353, 1979.
179. KELLER, M.D., LANESE, R.R., MITCHELL, R.I., ET AL. Respiratory illness in households using gas and electricity for cooking. II. Symptoms and objective findings. *Environ Res.* 19:504-515, 1976.
180. MILLER, R.W. Health implications of water contaminants in man. *Ann NY Acad Sci.* 298:557-60, 1978.
181. KAILEN, E.W. AND BROOKS, C.R. Systemic toxic reactions to soft plastic food containers. *Med Annals D C.* 32:1-8, 1963.
182. ZWEIDEINGER, R.A. Organic emissions from automobile interiors. *Chem Abstract.* 90:43297c, 1979 (abstract).
183. SUNDIN, B. Formaldehyde emission from particleboard and other building materials: a study from the Scandinavian countries. *Chem Abstracts.* 90:123390t, 1979 (abstract).
184. NATIONAL RESEARCH COUNCIL. *Indoor Pollutants.* Washington, DC: Natl. Academy Press, 1981 (2101 Constitution Ave. N.W., Washington, DC 20418;

ISBN 0-309-03188-5).
185. Niemela, R. and Vainio, H. Formaldehyde exposure in work and the general environment. Occurrence and possibilities for prevention. *Scand J Work Environ Health.* 7:95-100, 1981.
186. Dockhorn, R.J. and Smith, T.C. Use of a chemically defined hypoallergenic diet (Vivonex®) in the management of patients with suspected food allergy / intolerance. *Ann Allergy.* 47:264-266, 1981.
187. Rapp, D.J. *Allergies and the Hyperactive Child.* NY:Cornerstone Library, 1981.
188. Rapp, D.J. *Allergies and Your Family. NY: Sterling Pub, 1981.*
189. Coca, A.F. *Familial Nonreaginic Food Allergy.* (3rd ed). Springfield, Illinois: Charles C Thomas, 1953.
190. Corwin, A.H., Hamburger, M., and Dukes-Dubos, F.N. Bioassay of food allergens. I. Statistical examination of daily ranges of the human heart rate as influenced by individually incompatible foods. *Ann Allergy.* 19:1300-1311, 1961.
191. Corwin, A.H., Dukes-Dubos, F.N., and Hamburger, M. Bioassay of food allergens.II. Food-induced reactions of the heart rate of guinea pigs. *Ann Allergy.* 21:547-562, 1963.
192. May, C.D. Objective clinical and laboratory studies of immediate hypersensitivity reactions to foods in asthmatic children. *J Allergy Clin Immunol.* 58:500-515, 1976.
193. Schachter, S. and Singer, J.E. Cognitive, social, and physiological determinants of emotional state. *Psychol Review.* 69:379-399, 1962.
194. Scheier, M.F., Carver, C.S., and Gibbons, F.X. Self-directed attention, awareness of bodily states, and suggestibility. *J Personality Soc Psychol.* 37:1576-1588, 1979.
195. Pennebaker, J.W. and Skelton, J.A. Psychological parameters of physical symptoms. *Personality Soc Psychol Bull.* 4:524-530, 1978.
196. Rinkel, H.J. Inhalant allergy. Part I: The whealing response of the skin to serial dilution testing. *Ann Allergy.* 7:625-630, 650, 1949.
197. Hansel, F.K. Coseasonal intracutaneous therapy of hay fever. *J Allergy.* 12:457-469, 1941.
198. Lee, C.H., Williams, R.I., Binkley, E.L. Provocative testing and treatment for foods. *Arch Otolaryn.* 90:113-120, 1969.
199. Rinkel, H.J., Lee, C.H., Brown, D.W., et al. The diagnosis of food allergy. *Arch Otolaryn.* 79:71-79, 1964.
200. Miller, J.B. *Food Allergy: Provocative Testing and Injection Therapy.* Springfield, Illinois: Charles C Thomas, 1972.
201. Horrobin, D.F. Interactions between prostaglandins and calcium: the importance of bell-shaped dose-response curves. *Prostaglandins.* 14:667-77, 1977.
202. Boyd, I.A. and Pathak, C.L. The response of perfused frog hearts to minute quantities of acetylcholine, and the variation in sensitivity with season. *J Physiol.* 176:191-204, 1965.
203. Kolsch, E., Stumpf, R., and Weber, G. Low zone tolerance and suppressor T cells. *Transplant Rev.* 26:56-86, 1975.
204. Nossal, G.J.V., Ada, G.L., and Austin, C.M. Antigens in immunity. X. Induction of immunologic tolerance to Salmonella adelaide flagellin. *J Immunol.* 95:665-72, 1965.
205. Firer, M.A., Hosking, C.S., Hill, D.J. Effect of antigen load on development of milk antibodies in infants allergic to milk. *Br Med J.* 283:693-96,1981.
206. Kare, M.R., Schecter, P.J., Grassman, S.P., Roth, L.J. Direct pathway to the brain. *Science.* 163:952-53, 1969.
207. King, D.S. Food and chemical sensitivities can produce cognitive-emotional symptoms. In Miller, S.A. (ed.): *Nutrition and Behavior.* Philadelphia: Franklin Institute Press, 1981, pp 119-30.
208. Rapp, D.J. Food allergy treatment for hyperkinesis. *J Learning Disabilities.* 12:608-616, 1979.
209. Rapp, D.J. Re: Sublingual provocative food testing. *Ann Allergy.* 46:176, 1981 (letter).
210. Lehman, C.W. A double-blind study of sublingual provocative food testing: a study of its efficacy. *Ann Allergy.* 45:144-149, 1980.
211. Caplin, I. Committee on provocative food testing data does not strengthen validity of provocative subcutaneous food test technique. *Ann Allergy.* 32:53-55, 1974 (letter).
212. Breneman, J.C., Crook, W.C., Deamer, W., et al. Committee on provocative food testing. *Ann Allergy.* 31:375-383, 1973.
213. Breneman, J.C., Hurst, A., Heiner, D., et al. Final report of the Food Allergy Committee of the American College of Allergists on the clinical evaluation of sublingual provocative testing method for diagnosis of food allergy. *Ann Allergy.* 33:164-166, 1974.
214. Kailen, E.W. and Collier, R. "Relieving" therapy for antigen exposure. *JAMA.* 217:78, 1971 (letter).
215. King, D.S. The reliability and validity of intradermal and sublingual provocative testing: a critical analysis of the controlled

research, manuscript submitted for publication, 1982 (Langley-Porter Institute, San Francisco, Ca 94143).

216. FORMAN, R. A critique of evaluation studies of sublingual and intracutaneous provocative tests for food allergy. *Med Hypotheses.* 7:1019-27, 1981.

217. BRYAN, W.T.K. AND BRYAN, M.P. Cytotoxic reactions in the diagnosis of food allergy. *Otolaryngologic Clin N Amer.* 4:523-533, 1971.

218. BLACK, A.P. A new diagnostic method in allergic disease. *Pediatrics.* 17:716-723, 1956.

219. ULETT, G.A. AND PERRY, S.G. Cytotoxic testing and leukocyte increase as an index to food sensitivity. *Ann Allergy.* 33:23-32, 1974.

220. ULETT, G.A., ITIL, E., AND PERRY, S.G. Cytotoxic food testing in alcoholics. *Quant J Stud Alc.* 35:930-942, 1974.

221. LIEBERMAN, P., CRAWFORD, L., BJELLAND, J., ET AL. Controlled study of the cytotoxic food test. *JAMA.* 231:728-730, 1975.

222. CONNERS, C.K. *Food Additives and Hyperactive Children.* NY: Plenum Press, 1980, chapter 5, pp 69-85.

223. HOLOPAINEN, E., PALVA, T., STENBERG, P., ET AL. Cytotoxic leukocyte reaction. *Acta Otolaryngol.* 89:222-26, 1980.

224. MAYRON, L.W., KAPLAN, E., INTERLANDI, J., ET AL. Induction of antibody-mediated RBC lysis in food extracts. *Ann Allergy.* 38:323-338, 1977.

225. MAYRON, L.W. AND KAPLAN, E. The use of chromium-51 sodium chromate for the detection of food and chemical sensitivities. *Ann Allergy.* 40:94-99, 1978.

226. HIRSH, S.R., KALBFLEISCH, J.H., GOLBERT, T.M., ET AL. Rinkel injection therapy: a multicenter controlled study. *J Allergy Clin Immunol.* 68:133-55, 1981.

227. VAN METRE, T.E., ADKINSON, N.F., LICHTENSTEIN, L.M., ET AL. A controlled study of the effectiveness of the Rinkel method of immunotherapy for ragweed pollen hay fever. *J Allergy Clin Immunol.* 65:288-297, 1980.

228. MCGOVERN, J.J., GARDNER, R.W., BRENNEMAN, L.D., ET AL. Role of naturally occurring haptens in allergy. *Ann Allergy.* 47:123, 1981 (abstract).

229. TRUSS, C.O. Tissue injury induced by Candida Albicans: mental and neurological manifestations. *Orthomolec Psychiat.* 7:17-37, 1978.

230. TRUSS, C.O. Restoration of immunologic competence to Candida Albicans. *Orthomolec Psychiatry.* 9:287-301, 1980.

231. MILLER, J.B. A double-blind study of food extract injection therapy: a preliminary report. *Ann Allergy.* 38:185-191, 1977.

232. DADD, D.L., DADD, R.C., AND MCGOVERN, J.J. JR., *Nutritional Analysis System: A Physician's Manual for Evaluation of Therapeutic Diets with Special Emphasis on the Rotary Diversified Diet.* SF: Nutritional Research Publishing Co., 1980.

233. REA, W.J., PETERS, D.W., SMILEY, R.E., ET AL. Recurrent environmentally triggered thrombophlebitis: a five-year follow-up. *Ann Allergy.* 47:338-44, 1981.

234. PHILPOTT, W.H., KHALEELUDDIN, K., AND PHILPOTT, K.I. The role of addiction in the mental disease process. II. On the chemistry of addiction. *J Appl Nutr.* 32:20-36, 1980.

235. SPRINCE, H., PARKER, C.M., AND SMITH, G.G. Comparison of protection by 1-ascorbic acid, 1-cysteine, and adrenergic-blocking agents against acetaldehyde, acrolein, and formaldehyde toxicity: implications in smoking. *Agents Actions.* 9:407-414, 1979.

236. BERKOW, R. (ED). *The Merck Manual of Diagnosis and Therapy.* Rahway, NJ: Merck, Sharp and Dohme Research Laboratories, 1977, pp 1957-1979.

237. ROBERTS, S.A., REINHARDT, M.C., PAGANELLI, R., ET AL. Specific antigen exclusion and non-specific facilitation of antigen entry across the gut in rats allergic to food proteins. *Clin Exp Immunol.* 45:131-36, 1981.

238. JACKSON, P.G., LESSOF, M.H., BAKER, R.W.R., ET AL. Intestinal permeability in patients with eczema and food allergy. *Lancet.* 1:1285-1286, 1981.

239. HALL, R.H. *Food for Nought: The Decline in Nutrition.* NY: Vintage, 1974.

240. COHEN, F. Stress and bodily illness. *Psychiat Clin N Amer.* 4:269-286, 1981.